SMARTER STUDY

Strategies for High School & ATAR Success

A practical guide to effective study skills for strong academic performance

SHANÉE DUVEEN

NSW . YEARS 7-12 . HSC . FOR STUDENTS & PARENTS

Smarter Study: Strategies for High School & ATAR Success
A practical guide to effective study skills for
strong academic performance

First published in Australia by Shanee Duveen-Apostolou 2026
smarterstudysolutions.com.au

A catalogue record for this
book is available from the
National Library of Australia

ISBN: 978-1-7648026-0-4 (pbk)
ISBN: 978-1-7648026-1-1 (ebk)

Typesetting and design by Publicious Book Publishing
Published in collaboration with Publicious Book Publishing
www.publicious.com.au

For Alex and Bella

CONTENTS

Introduction i

A guide on how to use this book vii

PART 1: THE SET-UP FOR ACADEMIC SUCCESS

Chapter 1 – The giant leap – from primary school to high school 1

Chapter 2 – Mental preparation – some philosophy and expert views 13

Chapter 3 – Key principles for success – some home truths, and a myth 19

1. "There is no substitute for hard work." 19
2. Work smarter, not harder 20
3. The law of diminishing returns 21
4. The Pareto Principle (also known as the 80/20 rule) 22
5. Beware the enemy of good 23
6 The art of prioritising and time management 24
7. The myth of multi-tasking 32

Chapter 4 – Goal setting 35

Chapter 5 – Starting as you want to finish – applying theory to practice 39

1. Starting as you want to finish: At school 40
2. Starting as you want to finish: From school to home 46
3. Starting as you want to finish: The bigger picture 69

PART 2: HOW TO MAXIMISE YOUR GRADES

Chapter 6 – Overview of curriculum, subjects, and assessments 77

1. Curriculum and subjects...78
2. Academic streaming...78
3. Assessments and exams 80

Chapter 7 – How to conquer assessments and exams...93

1. WHAT are you being asked to DO? 96
2. WHAT do you need to KNOW? 98
3. WHAT is the QUESTION asking, and HOW do you answer?......................................99
4. WHAT is the MARKER looking for? 136

Chapter 8 – Assessment preparation and exam technique 185

1. Making notes..186
2. Studying and revision 197
3. Past exam papers..205
4. Reflecting on your holistic performance................221
5. Finishing touches and final preparation 226
6. English strategies for assessment preparation and exam techniques..........................229
7. Maths strategies for assessment preparation and exam techniques..270
8. General exam techniques 280

9. After the exam 286
10. A note on the additional demands of Years 11–12 295

PART 3: STRATEGIC PLANNING FOR ACADEMIC EXCELLENCE

Chapter 9 – Assessment and study planning....................303

1. Initial planning steps..304
2. Creating your assessment study plan 312
3. Additional planning TIPS for exam blocks, Year 12 Trials and HSC 323

Chapter 10 – Making strategic choices when the answer is not clear-cut 335

1. Subject choices and electives 335
2. Deciding whether to do 10 or more units for HSC..340
3. Subject acceleration..343
4. Deciding whether to get a tutor – or not..............346

PART 4: WRAPPING IT ALL UP

Chapter 11 – Key insights from top HSC achievers … and final thoughts 355

1. Insights from some top HSC achievers 356
2. Final thoughts and some universal truths 372

A very short conclusion 377
Acknowledgements..379
Bibliography and further reading..383
Notes 389
Index..391
About the Author 401

INTRODUCTION

When I started this, my daughter was finishing high school and my son was at university. I was not sure where this "document" would land up, or what form it would take, but I knew beyond doubt that there were many things my children, and our family, discovered the hard way during the high school journey that could be shared to benefit others. These were things I desperately wished someone had told us about, so we/they did not have to figure it all out, sometimes stumble, and unnecessarily re-invent the wheel. I wish we had been given a resource or manual that outlined what to expect, along with strategic and practical tips on how to prepare for and navigate the demands and nuances of high school to optimise success. I actually searched extensively for such a book – I scoured the bookshops, including the dedicated academic bookshops. I found heaps of subject-specific books; maths books full of practice questions, English study guides for set works, and books aimed at preparation for specific tests, such as NAPLAN, opportunity class, selective high school, and scholarship… but I could find NOTHING with overarching advice on how to approach secondary school in a way that would create the best overall results and signpost the path to achieving the highest possible ATAR. Of course there is an enormous amount of information out there, but you have to know where to look, how to discern what's important and relevant, and figure out how to decode and apply it once you find it, which can be challenging, time consuming, and confusing. This provided my inspiration for sharing learnings that our family, and specifically our children, have accumulated over the last ten years. I have combined this with valuable and relevant knowledge gleaned from the work and research of prominent academics and experts in the fields of mindset, academic achievement, and positive psychology, who I will reference along the way. I have also included key tips and strategies from some top-achieving students in the HSC. The goal of this book is to illuminate relevant information and provide clear guidelines, simple strategies, and practical frameworks which can be applied to elevate performance and facilitate academic success.

I guess you might be asking yourselves why you should take advice from a stranger who is neither a professional educator nor psychologist … and you would be right to ask this question. So let me tell you a bit about myself and my family. First, let me be transparent about all the things I am not: I am not a teacher, psychologist, life coach, or doctor of anything – although I did study an honours degree majoring in clinical psychology. Now that you know what I am "not", let me tell you what I "am".

First and foremost, I am a mother of two high-achieving children. I have been in the fortunate position to accompany them on their educational journey (at times probably a bit too much for their liking), and it has given me a front row seat to the choices they made, the academic frameworks they set up and the principles they applied in their pursuit of success. Through their experiences, we have **all** learnt very valuable (sometimes painful) lessons along the way, because let's be honest, the path through high school can sometimes feel treacherous and is seldom a smooth and linear upward trajectory. I say "all" because although their success is attributable solely to them, there has been an element of teamwork. Sometimes adult experience and perspective can be beneficial, and at the very least it creates support structures that assist them in doing what they need to do. Through this process, we have developed a significant knowledge base which would be applicable and appropriate for most students, helping them maximise their personal academic potential, whatever that may be. Academic success may look different for different students, but the principles and strategies presented here will help to equip every student with valuable tools allowing them to strive for *their own* personal best. For some students, the goal may be achieving top marks or working towards a specific ATAR, while for others it may be trying to improve their grades from the previous year. Applying the techniques outlined in this book will help move them towards that goal. It is important to acknowledge that success at high school cannot, and should not, be defined purely in academic terms, and there are many crucial aspects to making the most of the high school journey, including the development of leadership skills, co-curricular engagement, social connection, work–life balance, and very importantly, physical and emotional wellbeing. Although the focus of this book is geared unapologetically towards the achievement of academic outcomes, I would be remiss if I did not touch on these other important areas.

So, going back to the question of why you should be taking advice from me – I firmly believe that the proof of the pudding is in the eating (proverbially speaking), so the biggest testament to the effectiveness of the strategies and methods which we are sharing with you is the outcomes my own children and others around them have achieved through the (mostly) consistent and diligent application of these. Both my kids, although certainly capable and competent students, achieved outstanding success that they may not have initially thought possible. This was without doubt facilitated by employing the methods which will be outlined in this book.

By way of brief context, our son was awarded a place for Year 7 at a top performing selective school, although in the end we sent him to a boys private school. With lessons learnt from our son's journey, our daughter had the advantage of getting an earlier start with some of the strategies, which no doubt contributed to her educational success at an academically selective girls private school. They ultimately earned ATARs of 99.5 and 99.8 respectively. The point of me sharing my kids' academic success is to substantiate the effectiveness of the strategies they employed throughout. While both of them had significant academic goals, aiming for top results, I believe that none of their academic achievement happened by luck or intelligence alone, but rather by hard work, consistent effort, lots of practice, and the early implementation of systems and processes which set them up for success.

Aside from my role as a mother, I have a Bachelor of Business Science Honours degree in the special field of clinical psychology. After university, I pursued a high-pressure career in the corporate world. My first role was as a planner for a leading retailer and this formed a valuable foundation for everything that came afterwards. As a sidenote: you will notice throughout this book, I love to plan. I believe good planning is half the battle won. If you don't want to take my word for it, then perhaps take wisdom from Sir Winston Churchill who famously stated: "He who fails to plan is planning to fail."

After a few years in planning, I changed job and industry and eventually went on to head up a global centre of excellence, sitting on the board of directors of a global consultancy which was owned by a large multi-national company. My primary role was to head up a team, manage large-scale projects for some of the world's most valuable brands, and provide consulting and training to our clients and partners in approximately one hundred countries. Central to our business was a model of *psychological commitment*. While this role allowed me to travel extensively and provided a wonderful opportunity for professional challenge and growth, it also allowed me to focus on something I am passionate about – coaching and training others in the pursuit of excellence. The principles of excellence in achieving business success are not very different to those required for academic success at school, so my vast experience of coaching and training large and diverse groups of colleagues around the world created a solid foundation of skills and strategies which I have nuanced and applied to a high school context. In fact, many of these are valuable life skills.

My passion has always been, and remains, in the realm of helping people. As such, I have also spent recent years volunteering as a high school mentor, working with teens in a secondary school environment, through Raise Foundation, which is a wonderful not-for-profit youth mentoring organisation dedicated to providing support for teenagers in high schools across Australia. Through my experience with Raise, and that of my own children, I am acutely aware of the challenges young people are facing today, not the least of which is the pressure experienced at school, so if our learnings can help ease and improve the high school experience for even one student or family, it will be an achievement.

While there is usually no such thing as 'one size fits all', my belief is that the tools in this book are as close as it gets, having been developed and finetuned over many years with the benefit of my experience as a parent, a mentor, and a corporate executive. The strategies, techniques, and tips I share, supported by concrete examples and templates (where possible), are universally applicable, appropriate for a variety of ages, skill levels, and academic competencies. While there will be many competing ideas out there on how to optimise academic results, I believe it's critically important to go beyond theoretical approaches and look for information that has been tried and tested with observable success. My hope is that you will find this book to be not only informative but, more importantly, both practical and easy to implement immediately to strategically guide yourself (if you are a student) or your child (if you are a parent) to the success you/they deserve, through high school and beyond. I have also included some bonus pragmatic organisational and logistical tips to help smooth the process of transitioning to high school, as well as tips and strategies specific to Years 11–12.

My final note for any parents reading this – the most valuable gifts you can give your children are unconditional love and a good education. If you are reading this book, then you are likely already a committed parent, and this resource will hopefully provide further assistance in ensuring your child makes the most of their educational opportunities.

Thank you for reading ☺

A few quick notes:

- I have deliberately written in an informal and conversational style, so the content is relatable and accessible, so please forgive any grammatical liberties taken.
- You may notice when reading that I oscillate between addressing students and parents – this is deliberate because I believe this is a valuable resource for both – after all it is a team approach.
- Where syllabus changes are relevant, I have tried to accommodate for this, especially with English, which is the only compulsory subject for the HSC – but for up-to-date and accurate curriculum information, please refer directly to the NSW Education Standards Authority (NESA) website (https://www.nsw.gov.au/education-and-training/nesa).
- This study guide does not claim or aim to teach any subject-specific knowledge, and all references to subject content are included with the sole intention of illustrating a point or demonstrating a process.
- I am not affiliated in any way with any company, retailer, or business I mention, nor do I receive any compensation, financial or otherwise, from any of them. Any mention I make is purely based on my opinion and personal experience where I believe it may be helpful.

A GUIDE ON HOW TO USE THIS BOOK

It does not matter what year your child is in – this study guide is applicable and relevant to most students across all year groups, from the start of high school all the way until the end of Year 12. If your child is starting Year 7, they will get a head start, and if already in Year 12, the strategies are crucial, as they are instrumental in HSC success and can be applied immediately.

HOWEVER – not all parts are equally relevant for all stages of high school – for example, it wouldn't be helpful for a Year 7 student to read this entire book cover to cover in their first week of high school. It would be overwhelming and much of the content would be out of context. Equally, a Year 9 student would not need to read about the transition to high school. With that in mind, this is my recommended approach…

For students in late Year 6 or early Year 7

- I would recommend initially starting with **Part 1**, which provides guidelines to **set you up for success**, AND **Part 4**, which shares **inspirational and practical advice from high-achieving students**. This includes navigating the transition to high school, learning the frameworks and principles necessary for excellence, and understanding the practicalities of managing school in a way that sets you on the path to achieving your academic best.

- Once the first term of Year 7 gets underway, things will move quickly. Within a few weeks you will be over the initial hurdle and developing a routine. At this point you will most likely be introduced to the assessment programme and I recommend reading **Part 2**, which covers **assessments**, before you receive your first assessment notification. The first read-through will almost certainly feel overwhelming, but do not panic! I suggest that you treat the first read as a basic exercise in familiarising yourself with the content (even though some of it may seem too advanced and out of context). Once you have a broad understanding of how to understand, prepare for, and execute various assessments, you can then use this book as a reference tool or study guide throughout the entirety of high school, referring back to the detail in appropriate sections as they become relevant. Conquering assessments is the key to your grades, so this section will be an extremely valuable study resource in every year all the way up until the end of Year 12.
- Once you have got a handle on how to approach assessments, the next step would be covered in **Part 3**, which deals with **assessment planning and strategic decisions**. Students who are striving for top marks should read this section soon after Part 2, but be aware that some of the chapters in Part 3 might only come into play later or be relevant for certain scenarios. The recommendation would therefore be to do an initial read-through, paying close attention to the assessment planning (which will be relevant for all students), and making a mental note of the other topics which can be revisited later at the appropriate time.

<u>For students who are already in high school</u>

- If you are further ahead in high school at the time you acquire this book, there is no need to take a staged approach to reading it, because it is likely you would already have encountered assessments and been faced with many of the scenarios we discuss.

- The only chapter that would be irrelevant is Chapter 1 which deals with the transition to high school, but the rest of Part 1 (from Chapter 2) would be a necessary starting point because all subsequent chapters rely on this foundation.
- It is recommended to initially **read the book sequentially** in its entirety (from Chapter 2 onwards) to gain an overview. Thereafter, you can refer back to the detail in relevant sections as needed throughout high school.

For students in Years 11–12

- If you are picking this book up for the first time in Year 11 or 12, it's crucial that you understand the detail as soon as possible, applying new learnings immediately, with the goal of enhancing performance and maximising your HSC results and ATAR.

PART 1:
THE SET-UP FOR ACADEMIC SUCCESS

"You need to constantly be working towards your goals" (S.R. – 99.85 ATAR)

CHAPTER 1:
The giant leap – from primary school to high school

Regardless of whether your primary school was public or private, large or small, the jump from Year 6 to 7 is tremendous and universally challenging for students and parents alike. In fact, Term 1 of Year 7 can only be described as a shock to the system.

To be prepared for these changes, it's important to understand what they are:

- High schools are often larger with bigger grounds to navigate
- Cohorts can be larger – potentially nearing four hundred in a year group
- There is a shift from one primary teacher to different teachers for each subject
- Students move from class to class, rather than having a 'home' classroom
- Students may be assigned a senior peer support buddy
- Students may be assigned to pastoral care groups (mentor or tutor or house groups)
- There is a larger range of subjects to juggle
- Individual class sizes may be larger or smaller depending on subject and school
- There is usually a different mix of students in each class
- Students may be assigned a personal locker (depending on the school)
- There may be stricter uniform policies (depending on the school)
- School days may be longer

- Timetables are more formalised
- Lessons may differ in length (some schools have four subject periods per day of 80 mins each and some have seven periods of 45 mins each – or anything in between)
- Some subjects (such as history and geography) may be semesterised initially
- Students may have to choose electives (mostly from Year 8 onwards)
- There is often greater diversity within the student population
- There are often more clubs, cultural activities, and sports on offer
- There are more choices to be made, with more activities to juggle
- There is less parental involvement with students expected to be independent
- Teacher and school communications are addressed directly to students, and parents may not be included or informed
- Students have greater responsibility (for their behaviour, choices, and belongings)
- There may be formalised discipline, such as detention
- There may be a requirement to get a laptop, with strict policies around use of electronic devices at school, including phone bans
- The work gets harder, and the volume increases significantly
- Assessments and exams are more formalised
- Grades and marks are more formalised
- Deadlines are more rigid, with penalties for late or incomplete work
- There is greater focus on academic streaming
- There is usually a significant increase in the volume of homework assigned
- For better or worse (depending on the perspective), there is often greater focus on marks and results, with increased competition between students.

At this point, I am going to ask you to PAUSE and take a DEEP BREATH…

Although it seems overwhelming, especially if this is the first time you are experiencing the transition to high school, let me reassure you that things DO settle, and you will absolutely figure it all out. Remember that every student (and parent) starting this new phase will feel somewhat overwhelmed by the magnitude of change. Some students will see this as an exciting new chapter, because realistically, by the end of Year 6, many students have had enough and are ready to move on. Conversely, there will be students who feel apprehensive or anxious about leaving the safety of primary school, unsure of how they will manage the next stage.

As a parent, I felt a little anxious at the start (and at quite a few points along the way if I'm honest), and that is one of the key reasons for writing this. I hope that shedding light on some of the common challenges of high school and providing a pragmatic approach to tackling these successfully might make the journey seem more manageable and set you up to successfully achieve your goals.

I believe that forewarned is forearmed, so before addressing the academic aspects, let's briefly look at some practical ways to make the transition easier and create an optimal setting for success. A small caveat: many of these tips may seem like common sense, but in the interest of being truly practical, I am going to err on the side of including them.

TIP

Be informed!

Read ALL the information sent by the school in advance. In fact, read it as soon as you receive it, even if it's halfway through Year 6! In our busy lives we often put the information booklet at the bottom of the pile, or leave the email partially read in the inbox, because at the time of receipt it seems so far away that it borders on irrelevant. I too have fallen victim to the trap of forgetting the leaflet at the bottom of my in-tray or letting an email slide off the first screen of my inbox (essentially landing in email oblivion), mentally labelled as "a task for another day".

Bear in mind that schools often pre-empt common questions and provide valuable information in emails or brochures which are sent out well before the start of Year 7. Closer to the start of the first term, they will send out information regarding class placements and teachers, timetables, and instructions for the first day or week.

Many schools have weekly newsletters. Once school starts, be sure to read these as they often contain highly relevant information which, if missed, can make things tricky.

TIP

As you read the information:

- Immediately diarise key dates
- Make notes or create a to-do list of anything that needs to be arranged
- Set reminders for things that can only be done later.

TIP

Engage early, and a lot.

It's important that parents and students engage with all the opportunities provided by the school even before starting Year 7. This includes things like:

- Attending school tours to become familiar with the layout and facilities – schools usually publish a list of dates on their websites, for which you can register
- Looking at a school map
- Attending information sessions, orientation days, or activities
- Attending any pre-testing or assessment days (crucially important)
    - Pre-intake assessments may be used to stream classes at the start of Year 7
- Downloading appropriate school apps

- Joining parent social media groups (coordinators or parent reps may set these up)
- If you are given a school portal login, sign in and become familiar with its contents and navigation – these can be important sites for accessing information specific to the student (such as reports, medical information, timetables, teacher contacts, etc.)
- If given the opportunity, sign students up for clubs, sports or other co-curricular/extracurricular activities (a great way to meet people and it's fun!).

It's also beneficial to tap into non-school resources such as other parents and students – whether they be family, friends, or members of the community.

- Talk to parents and students already in high school – preferably (but not necessarily) the same high school. This is a critical way to benefit from the learnings of others and tap into existing resources.

Some students will be starting school with friends, or siblings in other years, but others may not know anyone. This can be a scary and isolating transition, so try to:

- Encourage them to get the names/contact details of one or two students who they have been introduced to at an orientation day or information session.
- Arrange social activities for students to connect ahead of time, so they have a familiar face when they begin school. This can help build connections and reduce some of the anxiety.

Why is it important to engage (early and a lot)?

- It's a practical way to get to know how the school operates
- It's a way for students and parents to meet and connect with each other
- It builds a sense of community
- Access to information creates a sense of internal control, which can help reduce the overwhelm factor

- Tip for parents: enthusiasm, support and investment in your child's school and education sends a message that education is valued and sets them up early for a positive and successful experience.

TIP

Be organised and prepared ahead of time.

Being organised in advance and feeling 'ready', reduces stress for students and parents. It creates a sense of control and makes for a positive start. There are a few things to arrange in this regard.

Uniforms (and haircuts)

Schools have different uniform and haircut requirements. Some have stricter policies and others are more relaxed, so it's important to read the policies in advance.

Many generic uniform items can be bought from mainstream retailers, but some items may need to be acquired through the official school uniform shop. In this case schools will usually advertise their purchasing options and shop hours through their formal channels or in the information pack. It may be possible to order online, or some schools might require an appointment for fittings. It's good to buy ahead of time, but not so far in advance that the student has a growth spurt before they start wearing the uniforms. Late in Term 4 or during the summer holidays usually works well.

It's recommended to label all clothing with your name, even if it's written with a laundry marker. You will be amazed how many items get lost at school, especially when changing for sport. Most unlabelled items vanish into a mysterious dimension of unknown location, never to be seen again.

Some schools have strict policies on haircuts. Be sure to find out and, if necessary, get a haircut the week before school starts.

Textbooks

Most schools issue textbook lists around November/December in preparation for the next school year. If ordering online through the school's preferred supplier, it's important to note the deadlines to ensure the books are delivered prior to Term 1.

While it's easy to order online via the school supplier, this can be more expensive, so check the academic bookstores – they generally have the booklists and sometimes offer a discount. They can also facilitate special orders and you can save on delivery.

You can also use or purchase second-hand textbooks. If you buy second-hand, ensure you get the same edition and version as the school prescribed list. Outdated editions may have different content no longer applicable to the syllabus, and different versions may have different page references, which makes it difficult to follow in class.

If you buy a second-hand, relevant edition, hard-copy textbook, there is often an activation code in the cover which grants access to an online version. This usually lasts one year from date of activation. If it has not yet been activated, that's great, but if the activation has expired, you can call the number provided within the book and buy a new activation code for a nominal fee. We did this numerous times and it worked well.

Many textbooks (particularly mathematics and science) have online versions, allowing you to opt for print, online, or both. The decision usually comes down to school recommendation and personal preference. My kids opted for both print and online as it suited their study style and they did not need to carry physical textbooks back and forth from school each day. They can be heavy!

Stationery

In high school, stationery lists are generally no longer issued. This is not a big deal – by now you can guess what will be needed: pens (blue and black), pencils, eraser, ruler, sharpener, highlighters, coloured pencils, glue, scissors, a maths set, and a scientific calculator. The school will usually suggest what calculators are recommended.

TIP(S)

If the school doesn't recommend a specific calculator, be sure to get a 'scientific calculator' approved by the relevant education authority – and if possible, get two!

In NSW, you will refer to the NSW Education Standards Authority (NESA), and the list is available on their website (educationstandards.nsw.edu.au). It's worth investing in an approved one at the start, as this will be used until the HSC. If possible, buy two – one for home and one for school. It's also a backup if one gets lost or the batteries run out. On maths and science exam days take BOTH to school, plus spare batteries!

***

As mentioned, one of the biggest adjustments in high school is moving from class to class, with multiple books for each subject, and homework for different subjects each day. It is therefore extremely important to be organised, as this reduces stress and facilitates goal achievement.

We used the following system which worked very successfully:

- We bought lightweight A4 document folders/wallets/pouches in various colours.
- Each subject was assigned a colour (e.g. maths – red, English – yellow).
- Each folder contained all books/notes for that subject (including exercise book, textbook, display folder and anything else that was required).
- The relevant folders were taken to school or home as needed, with everything for that subject kept together so nothing got lost or left behind.

- This makes things easier, saving time when moving between classes, especially if you have a locker – you grab the relevant subject folders containing everything you need so you don't risk leaving behind any necessary books or resources.
- We kept separate files for loose notes or work which could stay at home.
- Unless specified otherwise by the school, it's good to buy an A4 exercise book (with margins and ruled lines) for each subject (96 pages is usually a good size).
- 5mm grid books work well for maths.
- TIP: Label textbooks and exercise books (and ideally stationery) before the start of the school year. Definitely label your calculators, as these are expensive to replace.

Other random organisational <u>TIPS</u> and some cautionary notes…

- Provide parent and emergency contact details, and medical/allergy information to the school before the start of Year 7.
- Print the weekly timetable (this may be issued before Term 1 or once school commences) and put it on the fridge or other visible place, so it's easy to check what subjects you have each day, ensuring you have what you need for each lesson.
- Most schools have Week A and Week B rotations, which alternate with different schedules. Always keep track of the week you are in, so you have the correct items for each day. Note that not every term starts with Week A, so don't assume.
- If laptops are required, schools will communicate the process and will usually recommend a specific brand or model – this can differ between schools.
- A cautionary note regarding laptops: if you have a choice between models and brands, be sure to get one with good battery life, which will last the full school day. Carrying a charger to

school can be impractical and a battery that runs low before the end of the school day can cause immeasurable stress, especially for online tests/exams taking place later in the day.

With the above attended to, you should be well placed to start.

Here is a brief checklist to ensure you are ready for the first day of Year 7:

- ◊ I know what date school starts
- ◊ I know school start and finish times
- ◊ I know how I am getting to and from school
- ◊ I have a bus or train timetable (if relevant)
- ◊ I know where to go when I arrive at school on the first day
- ◊ I have my school uniform and a school bag
- ◊ My hair is cut according to school policy (if applicable)
- ◊ I have my textbooks and workbooks
- ◊ I have my stationery
- ◊ I have my laptop (if required)

A note for students (and parents):
As you walk through the gates on the first day of your high school journey, remember there are a lot of things you won't have encountered before. You might feel lost and like everyone is speaking a language you don't understand. This is a common experience for students. My best advice is:

Never be afraid to ask for help!

Older students are a great resource. They have been through it and know how it feels to start high school. Don't be shy to ask, most students will happily show you the way or answer questions. School leaders, such as prefects, are also great people to ask. They are usually awarded leadership roles because they have demonstrated a willingness and aptitude to be helpful. Teachers are also happy to assist students. They recognise how scary starting high school can be and they want to help students succeed.

It's much better to ask for help, whether it be a big or small issue, than to struggle silently. The sooner you ask, the sooner you will know what to do, and with each step, the adjustment to high school life will become more manageable.

A note of reassurance: with time, things will settle down and feel more under control. Every new thing you learn how to do will be one less unknown you have to conquer.

CHAPTER 2:
Mental preparation – some philosophy and expert views

Although you are now physically prepared to walk through the school gates on the first day of Year 7, there is some crucial mental preparation which, if you understand and implement early, can give you tremendous strategic advantage. That's what these next few chapters are devoted to....

When my son started high school, the head of Year 7 gave a speech. Most of that speech is now a distant memory, but one thing has stuck with me all these years. He said: "***Start as you want to finish.***" I believe there is great wisdom in those six deceptively simple words. This is what I take from it:

- It means actively **deciding** "how you want to finish". What is the end **goal**?
- It means **planning** the steps to achieve goals – you have six years, so there are many steps (they won't always be steps forward – it's ok to have the occasional side-step or even back-step)
- It means holding yourself to a **high standard** from the beginning
- It means holding yourself **accountable**
- It means being **consistent** in your **efforts** – it all adds up
- It means actively **managing** your **time** and **resources**
- It means being **resilient** and **perseverant** – not giving up, even when it's tough (it will sometimes be tough – this is guaranteed!)

- It means using the six years to build **skills**, **strategies** and develop **character**
- Quite simply – if you want to **finish well**, you should **start well**.

This doesn't mean kids who choose not to start well, or don't start well for other reasons, can't turn things around. Of course they can. Better late than never, which applies to most things in life. It just means the sooner you start working towards your goals and developing skills, the more time you will have to build to your desired finish.

***

I need to digress briefly to frame the importance of what follows, because starting off with the right attitude, mindset, and commitment undoubtedly sets you up for success.

This is a good time to reference Malcolm Gladwell, who is one of my favourite authors and has written books on a variety of topics. My top pick so far is *Outliers: The story of success* (2008). His writing is informed by substantial research and for full details I recommend you read his book. In the meantime, I am going to draw your attention to some of his key findings: In Chapter 2 – "The 10,000-Hour Rule" – he acknowledges the generally accepted notion that achievement results from some combination of innate talent and preparation, although there has been long-standing debate over the relative contribution of each in determining success.

Gladwell's book cites studies of expertise, which clearly demonstrate that beyond a certain baseline of talent, success and mastery is determined almost solely by the extent of practice and preparation. In other words, excellence requires practice. How much practice? Well, the answer is more exact than you might expect, and you've probably already guessed from the title of his chapter. The number is precisely 10,000 hours. Yes, ten thousand hours! Gladwell cites neurologist Daniel Levitin, who says "The emerging picture from such studies is that ten thousand hours of practice is required to achieve the level of mastery associated with being a world-class expert – in anything" (Levitin, D. 2006, as cited in Gladwell, M. 2008). He posits that "it takes the brain this long to assimilate all that it needs to know to achieve true mastery". Gladwell sums this up very eloquently in his conclusion that "Practice isn't the thing you do once you're good. It's the thing you do that makes you good" (2008).

So, how is this relevant for high school? First and foremost, it establishes that **practice and working consistently is critical for success**.

I did some rough calculations, and most students in Australia spend approximately 8,000 hours at school over the six years from Years 7–12 (accounting only for the academic school day). If you then add in time expected to complete homework, that amounts to approximately another 2,000 hours over the six years. This is a very rough estimate assuming less homework in the early years and more in the later years. This also likely excludes time spent on formal assessment and exam preparation to the extent it falls outside homework time, which it usually does.

So... it appears that school time plus basic homework time over six years lands up at around 10,000 hours. I am not sure if this is by design or coincidence, but it means that students who take school seriously, show up each day, pay attention in class, and work diligently to complete assigned work, will end up having completed at least 10,000 hours of school by the time they finish Year 12. This is a good baseline, because it means the system is set up to allow students to do enough hours to achieve some level of mastery in the academic domain. Having said that, realistically, schools can't and don't provide everything that a student needs to excel, so it's **how** students choose to work within the system and approach their studies that will be the differentiating factor in success. That's where I am hoping the tools I outline in this book will be helpful.

Before I get back to the practicalities, I would like to reference Carol Dweck, who authored an excellent book titled *Mindset* (2006). Dr Dweck is a world-renowned researcher and professor of psychology at Stanford University, and in her book she shares her decades-long, groundbreaking research and ideas on the power of mindset in achieving success.

I highly recommend her book! It's fascinating and easy to absorb, but for the sake of the task at hand I am going to briefly explain the key ideas and how they relate to academic achievement and success at school. Dr Dweck believes that innate ability and talent will only take you part of the way, and what truly determines success is whether you have a **fixed mindset** or a **growth mindset**. She defines a fixed mindset as a belief that "your qualities are carved in stone" and that your intelligence and abilities are fixed, innate and will stay static over time. A growth mindset is defined as "the belief that your basic qualities are things you can cultivate through your efforts" (Dweck, 2006). Dweck provides volumes of research and evidence on the impact of 'mindset' on motivation, learning, overcoming challenge, and achievement, including academic success. The evidence from her longitudinal studies demonstrated that during high school, kids with a fixed mindset tended to put in less effort and have declining grades over time, whereas those with a growth mindset had improved grades over time. Additionally, kids with a fixed mindset explained their poor grades by saying things like "I am the stupidest" or "I suck in math" or by blaming external factors such as teachers. Kids with a growth mindset, who may have been equally overwhelmed, adopted a different approach: "their response was to dig in and do what it takes" (Dweck, 2006). Very simply put (and for reasons explained fully in her book), Dr Dweck has found that students with a **fixed mindset** tend to put in less effort and under-achieve, while those with a **growth mindset**, who believe they can directly impact their outcomes, put in more effort and reap the rewards. The good news is that mindsets can be changed and taught. Often this starts with parents and teachers, and the language used in both praise and criticism.

A few things for parents to bear in mind:

- Praise and encouragement are important, and when praising your kids it's vital to resist the temptation to focus on **fixed aspects** such as intelligence or talent and rather focus on praising **growth aspects** like practice, diligence, perseverance, and other process-related efforts.
- Equally important is how you respond to disappointment. Resist the temptation to focus on fixed aspects of ability and avoid laying blame elsewhere. It is far more constructive to be truthful (in a kind way), helping to identify learnings that came from the disappointment and workshopping strategies to improve the next time.
- The word **yet** is very powerful. Try to end sentences with "yet", especially when giving constructive feedback or responding to a student's disappointment over something like a poor grade. "Yet" is infused with optimism and encouragement, implying that something can be worked towards and improved upon. It's not fixed, its growing. For example: "you haven't managed to achieve an A **yet**," implying that with effort, this can be achieved in future.
- It is very important for us parents to model the **growth mindset**, by implementing this for ourselves (which can sometimes be tricky). It means demonstrating that there is value in effort, rather than just absolute results.
- Remember – if the effort is there, the results will follow.

Dweck believes that "important achievements require a clear focus, all-out effort, and a bottomless trunk full of strategies" (Dweck, 2006).

Pulling this together in a way that hopefully makes sense – I believe there is great synergy between the ideas presented by Malcolm Gladwell and Carol Dweck. Both provide compelling evidence that innate talent and intelligence alone do not create success, and both attribute achievement to hard work and effort. Gladwell puts forward that you need **practice** and **preparation** to generate **success**, while Dweck believes that you need a **growth mindset** and **effort**. I believe the combination of these ideas is powerful and that students who are prepared to work hard, learn from mistakes, prepare well, and put in sustained effort from the start, are destined to finish well, maximise their potential, and succeed academically. These principles are equally applicable to students in all year groups, which means that whether you are starting out in Year 7 or doing your HSC, the application of this approach will pay great dividends. It is against this backdrop of knowledge that I aim to provide tried-and-tested tools and strategies to help guide and optimise those efforts to lead students to academic success.

CHAPTER 3:
Key principles for success – some home truths, and a myth

Following on from Chapter 2, which is largely broad and theoretical, let's talk in more concrete terms about what this means for HOW you approach high school. I believe there are some key guiding principles, which if understood and followed, will inevitably steer you towards success. It's important to outline these before we embark on the detail of navigating day-to-day high school, because these overarching principles apply to every year group and almost every facet of secondary school (and life). Establishing a solid foundation and applying these fundamental principles in a calculated, disciplined, and sustained way, will give you a head start and crucial advantage. They are skills which develop and become finetuned over time, underpinning academic success. It is indisputable that YOU have a great deal of control over your educational journey and what you choose to put in will determine your outcomes.

1. "There is no substitute for hard work."

Thomas Edison famously said it… Malcom Gladwell and Dr Carol Dweck both said it in their own ways … and even Dwayne "The Rock" Johnson states in social media accounts that "consistent, hard work gains success" (although he also said to always be humble and hungry – but that's another book). Now I am saying it too – **There is no substitute for hard work**. As with anything in life, the more you put in, the more you get out, and consistent hard work pays dividends. We see this in action every day, whether in the classroom, on the sporting field, or any other domain. Hard work, if consistent, allows you to learn, practise, and develop skills that translate to results. Very few people can excel without hard work and effort, no matter how intelligent or talented they are. Hard work is a key ingredient in building sustained academic excellence.

We have seen over and over again that top students tend to work harder than their peers. This is no coincidence. The earlier you recognise and adopt this strategy, the better your results will be over the long term.

That said – not all hard work is created equal, which leads to the second home truth …

2. Work smarter, not harder

At face value, this may seem like a direct contradiction to my assertions in the previous point, but I assure you it is not. If a baseline of **hard work** is good, you could be forgiven for assuming that **harder work** is even better. This is not necessarily the case. The strategic extension of **hard** work is actually **smart** work. This means more hours don't necessarily equate to more value – it's **what** you do with the hours that matters more. This is where **strategy**, knowing **WHAT** to do and **HOW** to do it to add value is critical, and this is what I aim to achieve throughout this book.

To illustrate this point, I will give you a common scenario: Imagine you have a science test approaching and you decide the best way to prepare is to study from the textbook, which is undoubtedly a great resource. You may devote a tremendous amount of time to reading and re-reading the textbook content until you have it memorised. You feel you know and understand it well. This sounds like a lot of hard work, and it is, BUT in fact, despite the hours, days, and maybe even weeks you have devoted to this preparation, if this is your only form of studying, there is a good chance you will not do as well as you'd like. The reason is not lack of time investment, commitment, or hard work, but rather lack of strategic preparation. Reading and understanding content is certainly very important for learning (as an initial step), but a smarter next step would be substituting some of the repeat reading (which is largely passive), with making notes, incorporating practice questions from the textbook, and doing past papers. Even if you spent less time overall, these types of activities would add tremendous value, and likely create better quality preparation.

Working smart therefore entails identifying the types or parts of work that will generate the best results and focusing on those. The best combination is working hard plus working smart, and if done properly, this can ultimately reduce the workload and save time. As we progress, I will point out clear and practical ways to work smarter.

3. The law of diminishing returns

When it comes to hard work, and even smart work, you need to be aware and cautious of the **law of diminishing returns**. Put simply, beyond a certain threshold, working harder (even if it's also smarter), will yield diminishing improvements. I will use numbers to illustrate this: Imagine your class has a history exam; you want to do well, and you work very hard and very smart. The results come back:

- You studied 22 hours and achieved a result of 96%
- Classmate A studied 14 hours and achieved 93%
- Classmate B studied 6 hours and achieved 75%

The extra eight hours classmate A studied compared to classmate B, translated to a significant increase in results. In this case, six hours of study was insufficient, and the extra work done by classmate A was clearly very beneficial, yielding an excellent return on time and effort investment (8 hrs of work led to an extra 18% in terms of results). However, you worked eight hours more than classmate A, and yet your result was only 3% higher. The biggest gain was won in the studying between 6 and 14 hours, while the gain between 14 and 22 hours was marginal. Was it worth it? Some students who are driven to achieve at the very top may believe the extra eight hours of studying was worth it to get 96%. For other students who may have been satisfied with 93%, the eight hours was perhaps not worth it, and could have been allocated elsewhere. It's important to note that the threshold is seldom obvious, and students often don't know where the point of diminishing returns lies, especially in early high school. It becomes a bit easier to gauge later as you gain experience and figure out the system. Additionally, the threshold can be different for each subject and unique to each student, depending on the effectiveness of their study strategies and their personal academic aspirations.

At the extreme, the law of diminishing returns comes into sharp focus in the HSC when we look at the top ATAR students. A disproportionate amount of time and effort is invested by the top students to get an ATAR just 0.05 higher. For these students who are driven to achieve at the highest level, they accept that a huge amount of work is required for an incremental gain, and for them, it is worth it.

4. The Pareto Principle (also known as the 80/20 rule)

Following on from working smarter and the law of diminishing returns, it's vital to understand the Pareto Principle (named after the Italian philosopher/economist who first came up with the concept). This is more commonly known as the 80/20 rule.

In simple terms: 80% of the results will come from 20% of the effort.

Again, we acknowledge that effort is crucial – it's the foundation, but in light of the previous three points (working hard, working smart, and recognising the law of diminishing returns), it reinforces the idea that a decent amount of hard work and solid effort will yield tremendous results, and if your efforts are smart, focused, strategic, and well executed, it can save a lot of time and stress. The trick is identifying **what** the most valuable 20% of the work is, because getting that right will generate 80% of your outcomes. Understanding this dynamic allows you to target your efforts to ensure you are prioritising the work that will garner the greatest gains (working smart) and also gives you a sense for where the line of diminishing returns sits (somewhere beyond the crucial 20%). This in turn helps with strategy, equipping you to make informed decisions about whether going the extra mile beyond that point is worth it for the smaller returns that will be generated, or whether your time might be better spent on something else. As we move forward, I will draw your attention to the specific tasks and types of work we have found to be in the most valuable 20%. I will also highlight tools and techniques which optimise this, providing guidance on where to invest the smartest effort to yield outstanding academic outcomes.

5. Beware the enemy of good

The French writer Voltaire wisely said: "the best is the enemy of the good." You will no doubt have heard many iterations of this sentiment, because it is an important one. The key idea is that striving for perfection oftentimes comes at the expense of making progress or getting things done. The trick is knowing when enough has been done to facilitate the desired outcome. This links closely with the previous two points of diminishing returns and the 80/20 principle, because spending an excessive amount of time on something for the sake of perfection, when it is not going to add appropriate value, amounts to time and effort wasted.

For example, imagine a scenario where you need to submit a landscape painting for a Year 7 art assignment. You have almost finished and are quite pleased with what you have created, but there are a few more finishing touches to get it 'perfect'. You spend two hours tinkering with the exact shade of blue for the sky and the shape of one of the clouds. After the two hours you may or may not feel you have reached your ideal, but in reality the extra time probably won't translate to extra marks, and if it does, it will likely be marginal. This is because these small details fell into the area of diminished returns and were not part of the most valuable 20% of effort. Aside from this, you need to ask yourself whether those valuable two hours could have been better spent. The answer will almost definitely be "yes". In those two hours you might have been able to progress, or even complete, other tasks in need of attention. I used this example because it's easy to grasp, but this same principle applies to literally everything, and it's a common trap that students fall into. It could be hours spent tweaking a Year 11 English homework paragraph or creating additional animations on a geography presentation. There will no doubt be times when the extra time and effort is necessary for something you deem very important, and as long as you are cognisant of the nature of diminishing returns and the 80/20 rule and make a deliberate decision that the extra effort is worthwhile, that's ok. The trick is knowing when it's time to stop and move on.

Creating the optimal outcome requires a fine balance between striving for excellence and making progress towards getting things done. Too much time beyond the optimal point can create inefficiency, additional stress, and create obstacles to progression, which can become an impediment to academic success.

6. The art of prioritising and time management

To be successful, all the previous points depend on the ability to **identify what is most important** – and to **prioritise** and **manage time** and **resources** effectively.

Time is finite and often constrained, so a key factor in success is developing tools to ensure that whatever time and resources you have available are spent on valuable activities which provide the biggest benefit. Conceptually, and practically, this is one of the most crucial determinants of academic success, and the ability to learn and master these skills early in high school will undoubtedly set you up to achieve your best.

Throughout high school (and life) you will have to juggle multiple competing demands. This is a certainty and one of the biggest shifts from primary school. It will mean, at times, balancing multiple assessment tasks or exams alongside homework, sport, co-curricular, leadership responsibilities, social activities, and personal or family commitments. There are also guaranteed to be some inconveniently timed curveballs.

The key to managing multiple commitments simultaneously is to be organised (I will go into more detail on this later) and being able to prioritise effectively. It's essential to recognise that not all tasks are equally important. If you are aware at the outset that there is a hierarchy, with some tasks being critically important, and some being irrelevant, this is the first step. If you understand what is most important, then it becomes clear how to apportion your effort and manage your time.

I will ask you to imagine a scenario. It's a Wednesday after school, and you have the following things to attend to: compulsory tennis practice on Wednesday afternoon, a vague plan to meet a friend for a milkshake, a maths exam on Thursday morning, homework for geography due on Monday, an English assessment due on Tuesday, and your friend has sent five texts asking for your opinion on which gaming console they should buy. You also need to clear spam messages from your email, and you want some time to relax and scroll Instagram. This is a straightforward example, and you will probably find it easy to discern the priorities, but let's put it in some sort of order:

1. Compulsory tennis practice (Wednesday)
2. Maths exam preparation (Thursday)

3. English assessment preparation (following Tuesday)
4. Geography homework (following Monday)
5. Meet a friend for a milkshake (sometime in the week)
6. Reply to the text messages (no deadline)
7. Clear spam emails (no deadline)
8. Scroll reels (no deadline)

There may be a debate around my proposed order, because it depends on what's most important in your hierarchy, and there may be personal or contextual issues at play, but for the purpose of the exercise, and assuming an academic focus, let's take the tasks and priorities at face value. Tennis practice is not negotiable as it's a compulsory school commitment at a fixed time. Next is the maths exam which is the following day, and an important task. Although geography homework is due before the English assessment, in terms of priority, the English is more important as it's a formal assessment counting towards your grade, therefore it should be attended to first. Homework is next, then social activity, text messages and lastly deleting spam and scrolling reels.

Taking this further, it's clear from this example that task importance is not the **only** consideration, and we also need to factor in another key dimension, which is time sensitivity, or urgency. The idea of considering both urgency and importance of tasks is not revolutionary and is often attributed to US President Dwight D. Eisenhower who observed that urgency does not necessarily equate to importance (1954).

Building on this historic principle, for our academic purposes it's critical to recognise that priorities don't necessarily follow the chronology of due dates, or the isolated importance of each task. Rather, effective prioritising depends on the ability to juggle and successfully balance tasks, with cognisance of both **time sensitivity and importance**.

Although prioritising based on **importance and urgency** may seem simple, it can be a complex exercise deciding the order in which you attend to tasks and the amount of time you spend on them, especially when the workload is overwhelming. It becomes even more intricate when you overlay the concepts of the 80/20 rule and diminishing returns. It can therefore be helpful to use a visual device to classify tasks based on relative importance and urgency. A visual representation clarifies what the priorities should be and therefore where time and focus should be directed first.

The **Smart Priority Grid** below is a study-focused decision framework designed to assist students prioritise tasks based on importance and urgency. To simplify, I have designated each quadrant a number, which I will reference throughout as we explore the significance of each.

Table 3.1 Smart Priority Grid

	URGENT	NON-URGENT
IMPORTANT	Quadrant 1 IMPORTANT + URGENT	Quadrant 2 IMPORTANT + NON-URGENT
UNIMPORTANT	Quadrant 3 UNIMPORTANT + URGENT	Quadrant 4 UNIMPORTANT + NON-URGENT

The fundamental idea is to focus on the important tasks first, prioritising them in order of urgency (Quadrants 1 and 2), and reduce time spent on things that are not important, even if they feel urgent (Quadrants 3 and 4).

Now, let's classify the tasks in our previous example according to where they might fall in the Smart Priority Grid quadrants:

1. Tennis practice (Quadrant 1 – important and urgent)
2. Maths exam preparation (Quadrant 1 – very important with a pressing deadline)
3. English assessment (Quadrant 2 – very important, but less urgent, you have time)
4. Geography homework (Quadrant 2 – important, but less urgent, you have time)
5. Meet a friend for a milkshake (Quadrant 2 or 4 depending on how important this is in your personal hierarchy – possibly socially important, but there is no deadline)
6. Reply to the texts (Quadrant 3 – not important but feels urgent/pressure to reply)
7. Clear spam emails (Quadrant 4 – neither important, nor urgent)
8. Watch reels (Quadrant 4 – neither important, nor urgent)

The ideal way to effectively manage a high school workload when working smart to achieve academic success, is to proactively spend as much active working time and effort as possible in Quadrant 2, which is productive and quality focused. This is where most good value planning and preparation takes place and where your time and effort will have the biggest payoff because you are devoting resources to tasks that are important, and don't need to be rushed. You would ideally minimise the amount of time spent in Quadrant 1, as this can feel like crisis management, dealing with work that is both important and urgent, requiring immediate attention. Tasks in this quadrant can't be avoided or delayed and can become stressful. It's unrealistic to expect you can completely avoid Quadrant 1 but keeping it to a practical minimum is optimal. You want to reduce Quadrant 3, which hassles you with urgency but doesn't add much value because the work is not important. You should have a goal to eliminate time spent in Quadrant 4, which is a waste of time, and basically consists of procrastination, consuming valuable time with things that are neither important nor urgent. Quadrant 4 is fine during holidays or rest periods, but keep in mind we are talking about this in the context of managing schoolwork.

In summary – Quadrant 2 has the biggest return on time and effort and constitutes the smartest working. Quadrant 1 is next, although doesn't provide quite the same gains, and is more stressful. The payback on time and effort is very low in Quadrant 3, either because you are working harder, not smarter, and perhaps spending time on the wrong type of work, or because you are focusing on things unrelated to work, which are mere distraction. Quadrant 4 offers no return unless wasting time is the goal.

Being able to prioritise your commitments and tasks in order of importance and urgency helps you to determine the **order** in which you attend to tasks, and the **amount and quality of time** you devote to them. This process is invaluable in ensuring you manage your time, allocating it to the right activities to achieve the best academic gains. Prioritising is fundamental to working smarter and facilitates focus on the most valuable 20% of effort, which will create the biggest gains.

Let's now look at practical steps to implement this …

TIP

FIRST make a comprehensive to-do list with everything (including 'time wasters')!

Being able to prioritise starts with a complete picture of **everything** you need to do. You can't prioritise it if you don't know you have to do it. Your list should include everything requiring your attention, down to the smallest task and, if possible, break things down to subtasks. Only once you have a complete list can you begin to assess your task landscape and work out the priorities.

This can be done on paper, electronically or in an app. Many apps have the capability to create lists (and sub-lists), synch to an online calendar, set reminders or even repeats for tasks which are likely to recur. This is an area where technology can be helpful.

TIP

Use a visual template to plot your to-do list by urgency and importance.

In practical terms, it's not always straightforward to identify priorities, although this becomes easier with practice. When you are initially deciding what to tackle first, it's helpful to use a blank grid (similar to the example introduced earlier), and allocate tasks to each quadrant. This is a valuable organisational tool which can assist in clarity of thought. Writing things down and creating a visual representation of your to-do list prompts you to engage with the notions of importance and urgency, thereby naturally guiding you towards good priority setting. It also reduces the stress and cognitive load of trying to weigh it all up in your head, which can feel intimidating and overwhelming. Over time and with practice, you probably won't need to use the physical grid as you build prioritising and time management skills.

This is a simple template you can use to establish priorities and manage time:

Table 3.2 Smart Priority Grid Template

	URGENT	NON-URGENT
IMPORTANT	Quadrant 1: important + urgent tasks TOP PRIORITY – DO NOW	Quadrant 2: important + non-urgent tasks HIGH PRIORITY – SCHEDULE IT
UNIMPORTANT	Quadrant 3: unimportant + urgent tasks LOWER Priority – MINIMISE IT	Quadrant 4: unimportant + non-urgent tasks LOWEST Priority – DROP IT

Once you have populated the Smart Priority Grid with the items on your task list, try and attend to them in order as far as possible:

- Focus on completing Quadrant 1 tasks first
- Then schedule in appropriate time for the Quadrant 2 tasks
- Try not to schedule Quadrant 2 tasks too close to deadlines or they are at risk of becoming Quadrant 1 tasks later
- Quadrant 3 tasks come next and should be done as swiftly as possible
- Try and avoid Quadrant 4 altogether, but if that's not possible, ensure you leave these items until all the important tasks have been done.

> **TIP**
>
> **Manage and allocate time and effort in proportion to the *importance* of the task – while also being mindful of the traps of diminishing returns and striving for perfection.**

The urgency classification can determine how soon you should attend to something, but the importance should determine the amount and quality of time and effort dedicated to a task. Invariably the important tasks will fall into the most valuable 20% of effort, with the greatest benefit, so it's worth prioritising and managing your time with this in mind.

7. The myth of multi-tasking

When you find yourself in a situation with multiple tasks to attend to (especially if they feel urgent), it can be tempting to try and work on them concurrently, or if you are feeling stressed and panicky, you may switch back and forth trying to make progress on as many as you can. In this situation, **BEWARE – multi-tasking does not work**. Take care not to fall into the trap of believing you are the exception.

There has been significant research (going back decades) around the controversial topic of multi-tasking. The dominant conclusion is that it's almost impossible for the brain to attend to two or more things at the same time. Even if it seems like two things are being done concurrently, they are actually being alternated in quick succession. This means every time your brain switches between tasks, you introduce inefficiency and gaps in attention, which lead to missed information, more mistakes, and slower processes. Most of this happens in the background, below your conscious awareness. At best, multi-tasking slows you down, but at worst it reduces quality, productivity, and information retention, ultimately leading to lower achievement. Multi-tasking can work if you are doing simple, automatic tasks which don't require much focus, such as going for a walk while talking on the phone, or exercising while listening to music, but as soon as tasks increase in complexity, multi-tasking creates issues in speed and accuracy. This opposes the principles of good time management, achieving the opposite effect, and thus has important implications in terms of the Smart Priority Grid. As a rule, for important tasks in Quadrants 1 and 2, you should aim to work on one task at a time without interruption or switching, as these tasks require focused attention to achieve quality outcomes. Even for Quadrant 3, where you need to get unimportant things done urgently, it's better to work through the list one item at a time efficiently. For Quadrant 4, tasks that are neither important nor urgent, interruption is likely less of an issue, and if we are honest, it's the Quadrant 4 items that are most likely the culprits, distracting you from work in the other quadrants.

TIP

For schoolwork, studying, or doing anything requiring focus:
- **Close your email or any group chats**
- **Log out of social media sites**
- **Put your smart phone away – FAR AWAY**
- **Turn your music off**
- **Avoid switching between tasks (even other important ones).**

All of these are detrimental to focus and productivity. For example, if you are busy drafting an English essay and interrupt yourself to quickly attend to an email that popped into your inbox, you will slow your process down. Even if it only takes you a minute, the cost is greater than you realise, compromising productivity and quality, even if you are not conscious of it. Understanding this is a big part of staying in Quadrants 1 and 2, and minimising distractions and time wasters that tempt you from Quadrants 3 and 4. This is intuitively easy to grasp with obvious distractions, but it's equally important to remember that the same rule applies even if you are multi-tasking between two different, and important, Quadrant 2 activities. Part of working smart is to **focus on one thing at a time** in dedicated blocks of time.

***

To sum up this important chapter – working hard and smart, navigating the point of diminishing returns, applying the 80/20 principle, resisting the temptation to make things perfect at the expense of getting things done, prioritising, time management, and eliminating multi-tasking, are all vital for academic success throughout high school. I will therefore reference these principles often and give you guidance on HOW to apply them in a practical way for optimal success.

CHAPTER 4:
Goal setting

The topic of goal setting was originally bundled into Chapter 3 but given that effective goal setting is vital to success, I believe it deserves its own section. Goals are cyclical in that setting them is the crucial starting point, achieving them is the end point, and once achieved, the next goal is set in motion, as we move onwards and upwards. Most often in life, and at school, we are juggling multiple concurrent goals, at various stages of completion, with varying magnitude and importance. These goals should ideally be set and executed with the broad principles of the previous chapter in mind as they go hand in hand. Essentially, we must work out what's most valuable and important for us to pursue, and consequently decide how much, and what type of effort we devote to our competing endeavours. We will then be able to effectively prioritise and manage our time and resources accordingly. To manage this, we first need to take a step back and understand how to set goals and evaluate achievement.

At the start of each high school year, my kids always wrote down some personal academic goals for the year. This was never formal or fancy, and often written on a scrap of paper, sometimes taped to the wall above their desks. While this sounds simple and trivial, identifying and committing to goals in a tangible, visible way is a tremendously powerful exercise for students for many reasons:

- It reveals what's **important to them**
- It declares their **intentions** (what they want to focus on)

- It creates a starting point to develop a **plan** or steps towards the **goal**
- It clarifies what their **priorities** will be
- ... which creates a guide for **time management** and **effort allocation**
- Committing to a **goal** in writing is more likely to result in follow-through
- ... and primes the brain to work towards it in the background
- It fosters intrinsic **motivation** as goals are self-defined
- ... intrinsic motivation is superior to extrinsic reward (or punishment)
- It creates a springboard to discuss **HOW** they might achieve the goals
- ... and guides parents on how best to support their pursuit of them
- It gives them something concrete against which to measure their **progress**
- If goals are achieved, it creates a valuable internal dialogue, and sense of **mastery**
- If goals aren't achieved, it's an opportunity to **reflect, refine, and plan** forward
- It becomes the basis for **future goals**.

When setting goals, it's important to ensure they are '**SMART**' – **specific, measurable, achievable, relevant, and time-based**. You may have already encountered the principles of 'SMART' goals, as this is a well-established model, first introduced by George T. Doran in an article he wrote for the *Journal of Management Review* (1981). Since then, various versions have been widely adopted and applied in business, education and other domains. The 'SMART' principle is important for effective goal planning, so I will briefly explain. Goals should be:

<u>S</u>pecific – the desired outcome should be precisely defined and clearly understood rather than vague. The more concrete the goal, the greater likelihood of achievement.

Measurable – the outcome must be quantifiable and measurable so you can track progress and be sure of attainment.
Achievable – the goal should be realistic, or at least within the realm of possibility, otherwise it will be a futile and demoralising exercise.
Relevant – it needs to matter and have relevance to the bigger picture. A goal that has no meaning is pointless.
Time-based – there must be a reasonable deadline for completion, or this can undermine motivation to prioritise.

Let's look at an example:

A student decides they would like to excel in their Year 10 mathematics course. This is a great goal, but it's vague and subjective, making it difficult to work towards, and more challenging to evaluate whether they have achieved the goal. To assist with goal setting they employ the SMART criteria, which might look something like this:

Specific – I would like to achieve an 'A' (85%) in maths on my end-of-year report. (This is very precise and the student will know with certainty whether they have met the goal. They will either be awarded an A, or not).
Measurable – I need an average of 85% or higher across the four assessment tasks. (This is quantifiable and easily measured. It's also easy to track progress to make the necessary adjustments, because if the student gets 85% in the first task, they know they are on track, and if they get less, they need to increase their effort to score higher than 85% on future tasks).
Achievable – I know this is realistic and possible because I got a B+ last year with very little effort. (This is clearly within the student's capability.)
Relevant – This is important because I want to qualify for Extension 1 Maths in Year 11. (The goal matters, because it is linked to a higher aspiration.)
Time – The deadline is the end of the academic year. (This has a very clear end point, by which time the goal will have definitively been achieved or not achieved.)

Defining goals in these terms makes it clear to the student what they want to achieve and allows them to plan and track progress accordingly. A goal that contains the SMART considerations sets the student up for greater effectiveness and success.

This goal setting model can be applied to almost anything, and when linked with the principles in Chapter 3, creates a powerful framework for excellence. When goals are clearly defined it's easier to identify priorities, establish which activities will create the greatest return on effort in terms of the Pareto Principle, and facilitate effective planning and time management. SMART goals are the foundation of smart work and time spent on goal setting is an essential Quadrant 2 activity.

Along the high school journey, there will be a multiplicity of goals in play. Some will be short term, some long term, some will be linked to bigger life goals, while some will involve just getting through the week. If we take on board the wisdom of "starting as you want to finish", defining your **end goal** at the start gives you a significant advantage, because from the outset you have a clear vision of where you want to land up, and that paves the path to success.

CHAPTER 5:
Starting as you want to finish – applying theory to practice

The previous chapters have set the stage and established the foundations, focusing on getting physically ready and mentally primed to approach secondary school in a way that creates optimum advantage. By now, you have hopefully bought into the benefits of hard and smart work, time management, prioritising and goal setting, and you have the best intentions to commit to the process, start on the right foot, and sustain the application of these principles all the way through to the end of the HSC…

… But, as we have established, high school involves change, and it's not unusual to feel overwhelmed at the start, making it difficult to apply the principles we have discussed. There is a consequent danger that the ideals of the success framework might get sidelined in the interests of survival in a new terrain. This is to be expected to some extent, but I assure you that as time goes on, you will figure out how things work, and in the meantime, if you can hold onto the key foundations of consistency, effort, self-discipline, goal setting, and effective prioritising and time management, this will stand you in excellent stead to conquer the system and achieve success.

It can also be difficult to translate this theoretical understanding into practice in relation to the daily demands of school. Although you may comprehend **WHAT** should ideally happen, it can be challenging to know exactly **HOW** to apply it. A key piece of advice is to pay attention to the details and employ a diligent, disciplined approach to some of the seemingly mundane, routine aspects of high school. The difference between achieving top results and mediocre results can boil down to the small day-to-day choices made by students both inside and outside the classroom.

The purpose of this chapter is therefore to tackle some of the concrete, tangible facets of secondary school – things that students are likely to encounter every day. These are often overlooked or underestimated and oftentimes students are left to figure it out on their own, but understanding **how** to approach these is crucial in navigating the system effectively to optimise success. Accordingly, I will outline some tips and practical strategies which align with the key guiding principles to create advantage. If established at the start and applied consistently, these will complement the skills outlined in the previous chapter, guiding you towards an academically successful finish.

1. Starting as you want to finish: At school

In class – pay attention, participate, ask questions, and take notes!

Paying attention in class is foundational to learning and critical for working smart. This is the best opportunity to absorb and process new information. It's also the ideal time to ask questions and seek clarification. In terms of the 80/20 rule, listening in class and maximising learning undoubtedly forms part of the most valuable 20%, offering a significant return on investment. You have to be in the class, so you might as well get the best value out of that time. If a concept is well understood in class, and classwork is completed properly, everything else builds from there. In fact, maximising classroom learning is essential, because if this is lacking or incomplete, everything afterwards will be built on a shaky foundation, containing gaps which inevitably compromise performance and impact results. Hours spent productively at school add up, and when you are trying to accumulate 10,000 hours, every hour counts.

TIP

Never be afraid to ask a question in class!

Often students avoid asking questions because they fear being looked down upon or appearing "silly", but I can almost guarantee there are other students in the class with similar questions who are also reluctant to ask and will be grateful to you for having asked on their behalf. Asking questions helps everyone. A secondary gain is that teachers appreciate those who ask questions because it indicates student engagement and facilitates a dynamic learning environment.

TIP

Actively participate in class and contribute to class discussion.

This is incredibly valuable because it creates active learning, which is more effective than passive learning. This also forms an important part of the valuable 20%, facilitating better quality learning and depth of understanding. It further gives teachers a better opportunity to get to know you as a student, which is advantageous.

TIP

Take notes in class.

Class notes form the foundation of study material as teachers generally teach the content that they are most likely to examine, so making notes during class will be helpful later. This sounds obvious, but you would be surprised how many students either don't recognise, or underestimate this. These study notes fall into

Quadrant 2, part of the essential 20% of work. They are beneficial and will save you substantial time later. (I will come back to this in a later chapter.)

The extent of optimal note-taking can differ by subject, student, or teacher, but a good general rule is to jot down notes for anything significant you will be expected to know and remember later. Here are some examples of useful things to make notes on **during class** (while it is fresh in your mind):

- A new or difficult concept
- A definition
- Key points of a class discussion
- A worked example (this is especially helpful for maths)
- Anything you know is going to be important that you may forget later
- Information related to syllabus dot points (key aspects of the prescribed curriculum)
- Anything the teacher emphasises or hints will be in an exam
- Information in a slide presentation accompanying the lesson

This list is not exhaustive and it's always best to use common sense. It's also not helpful to write down every word the teacher says, because multi-tasking is not ideal, and it's important to be focusing on the lesson itself. There is skill required in discerning what is beneficial to note down, but this skill will develop over time with practice.

Most study experts recommend that every afternoon students spend some independent time on each subject consolidating key learnings for the day and making notes accordingly. I agree that this is a helpful strategy and a wonderful ideal. For students that can manage this, I would strongly encourage it. However, realistically for most students this just doesn't happen. Time after school is scarce, especially if there are extracurricular activities, and the remaining time is usually devoted to deadline-driven homework or assessments. While creating study notes, working ahead, and consolidating the day's learning is valuable and would fall into the desirable Quadrant 2 category, it is often not realistic, especially in the early years, as these quality-driven tasks tend to be non-urgent, falling lower in the priority list to make space for more urgent tasks. I will address this again later, because the demands and recommendations are different in Years 11–12.

> **TIP**
> **Never leave a classroom with uncertainty.**

This applies to concepts taught, homework expectations, or any enquiries relating to a subject or lesson. Although this sounds logical, I can't express the number of times my children came home (especially in Year 7) unsure of something – whether this was regarding what questions to complete for homework, when something was due, or not understanding an example from class. In my experience, it creates a lot of stress and takes unnecessary time to resolve later at home. There are always ways to get clarification, including emailing the teacher or calling a friend, but it adds time to the process without necessarily adding value, and is therefore an example of not working smart. Working smart is resolving things as they arise, in the most expedient manner.

> **TIP**
> **Don't get distracted with other things during class time.**

The reason I mention this is because you would be shocked to know how many students spend class time playing video games on their laptops or chatting to their friends – even in high-performing academic environments.

I am sure I don't need to explain why this is not recommended. This is a prime example of a Quadrant 4 activity; a pure waste of time which impedes learning and understanding. The price is usually paid later.

Try not to miss class...and if you do, be sure to catch up promptly

Try not to miss classes, because this is high-value learning time. There will, however, be occasions when this is unavoidable, for example, you may be unwell, have a scheduled commitment, or a school excursion. If you miss a class for any reason, it's advisable to email the teacher to let them know and find out what work you need to catch up on – either before the class if it's a planned absence, or soon after the class if it's an unexpected absence (ideally the same day). This has multiple benefits:

- The teacher will appreciate the courtesy of being notified.
- It creates a good impression and signals to teachers that you are proactive, have a good work ethic and take your studies seriously.
- You will know what work has been missed that you need to catch up on.
- You will know what homework has been assigned that needs to be completed.
- You will minimise learning gaps.

It's also a good idea to ask one or two people in your class if they can send you notes or handouts as soon as possible, before it falls off the radar.

In some cases, if feasible, teachers may allow you to attend class remotely via Zoom so you can avoid missing work. Not all schools accommodate this, but it's worth asking because it reduces the burden of catching up.

Being proactive in catching up work will reduce the potential for academic gaps, which could be detrimental later. Be aware that if you don't take the initiative to chase down the information, people don't tend to bring it to you, and you may risk missing out.

Communicate with teachers directly

It can be tempting in early high school for parents to email or call teachers on behalf of students. This might happen because:

- It's what happened in primary school, and it's a habit
- Students may feel shy to reach out to teachers
- Parents may have anxiety about taking a step back.

There will, of course, be legitimate instances where it is entirely appropriate for a parent to contact a teacher directly, but for issues which fall within the domain of the student, it's important for students to take responsibility for the communication. This is an important step forward for students (and parents), which facilitates growth as:

- Students gain confidence by speaking for themselves.
- Students learn to take responsibility for their education.
- They learn valuable lessons in professional communication.
- It enhances the learning journey when students seek clarification.
- Appropriate communication engenders mutual respect between students and teachers and, in general, teachers prefer to manage student queries directly rather than via parents or carers.
- It becomes easier and more natural with practice, which will be beneficial in later years as parents are increasingly left out of the loop.

Take on board ALL feedback

Feedback is a certainty in high school. This will be provided frequently in many forms. It may be formal or informal, verbal or written. It may be provided for classwork, homework, assessments, or on school reports. All forms of feedback are valuable and delivered with the goal of helping you to improve. Students with a growth mindset, who absorb and actively apply feedback will derive tremendous academic benefit.

Behave well. Consistently. From the start.

Behaving well facilitates an optimal learning opportunity, while disruption derails learning, but there is also a more subtle implication which can come into play later.

Teachers form an impression of students based on their behaviour, including class participation. These impressions can help or hinder you down the track without you being aware. No matter how objective a system is designed to be, teachers are human and consequently there is always some element of subjectivity in a teacher's assessment of a student. It's unavoidable and can have a subliminal impact when writing student academic reports or marking assessments. If something comes down to a 50/50 judgement call (like a decision to award or deduct a mark in an ambiguous situation – which happens more often than you think), students with a track record of exemplary behaviour and demonstrated consistent effort will usually be given the benefit of the doubt. Unfortunately, the opposite is also true.

The behaviour rule extends beyond the classroom, and in senior years a student's behavioural record is likely to be instrumental in decisions regarding the appointment of leadership roles.

It is also becoming commonplace for schools to place greater emphasis on learning dispositions, with behavioural and social criteria. This is sometimes incorporated into academic reporting, forming part of a student's official school record. This will be important beyond high school when applying for tertiary study and jobs.

2. Starting as you want to finish: From school to home.

Do your homework!

One of the greatest differences between primary and secondary school is homework. The escalation in volume and complexity of homework is considerable, and significantly greater demands are made on students in terms of time investment and cognitive load.

In primary school the emphasis is largely on straightforward consolidation of daily learning, and is mostly concentrated on English and maths, consisting of foundational skill-building tasks such as spelling, grammar, handwriting, learning multiplication tables, basic arithmetic, doing worksheets, and preparing the odd report or speech. This is assigned predictably on a regular daily or weekly basis, and because there is one teacher who oversees everything, the volume of work and difficulty of the content is moderated and doled out in manageable chunks. The deadlines are generally not onerous, and there is usually not much other schoolwork to be juggled. Students hand in homework to the class teacher, who marks it, hopefully provides some feedback and returns it. All of this is entirely appropriate for primary-aged children.

In contrast, from literally the first week of high school, students are thrown into the deep end, left to 'sink or swim' in a sea of homework, and many kids are not prepared for how to navigate or manage the process effectively. Year 7 homework can initially be a huge shock!

These are some of the ways homework in Year 7 differs from Year 6:

- There are ten or eleven subjects with different teachers, all assigning homework at their own discretion on an irregular schedule – this can lead to homework overload.
- Teachers don't mark all the homework and sometimes don't check if it's been done – which means there may be no follow-up or prompting.
- Homework must be juggled with other schoolwork and therefore requires planning.
- It is expected (but not always communicated) that students complete classwork in their own time in addition to homework.
- There is often a guideline that Year 7 students should only do one hour of homework per afternoon, with the time increasing by approximately 30 minutes each year – however this is almost always unrealistic, and homework invariably takes longer.
- Deadlines are variable, strict, and there may be quantifiable penalties for incomplete or late work (often reflected in a poor effort score on the report for that subject).

- Maths homework tends to get assigned EVERY lesson, and depending on the school structure and timetable, this can be every day – and is generally a lot.
- Students are expected to manage their homework process independently.
- The complexity of the homework increases and becomes more application based.

Homework is a controversial topic, with some believing it has value while others believe it adds excess stress and pressure. On the plus side, there are benefits to be derived from forced revision, consolidation, and practice, and it helps build towards your 10,000 hours. On the minus side, homework can sometimes be time consuming without adding proportionate value. I am not going to make a judgement or weigh in on this debate, because in my experience, most schools subscribe to the value proposition, meaning homework is here to stay. Given the status quo, it's important to find ways to work within the system to optimise the benefit and outcomes. With this in mind, we quickly learnt that we needed a system for managing homework effectively!

Here are some key <u>TIPS</u>...

- Ensure classwork is up to date as a priority, even if it means completing it independently at home. Teachers may not be explicit about this, but it's assumed that students will stay up to date. Falling behind and leaving work incomplete leads to learning gaps, which compound quickly and compromise results later.
- Keep up to date with assigned homework. Teachers may not check, and it's left to the students to self-manage. Diligent students who take responsibility for their learning and complete assigned homework will fare better in high school. They accumulate and consolidate knowledge and develop and practise skills creating greater academic competencies. They are actively

working towards accruing their 10,000 hours to achieve mastery. It is possible for students to neglect homework, and for a while this may escape the notice of teachers with no immediate consequences, but eventually this creates an academic disadvantage which negatively compounds over time.

- The homework flow can be erratic. Some days or weeks there may be an avalanche of work, and other times a trickle. In weeks where assigned work is sparse, it can be tempting to enjoy the reprieve and neglect homework, thinking you will deal with it later, but this can backfire when the next wave hits. It's best to attend to the work as it is allocated to avoid a backlog, which creates undue stress. It's even better if you can use some of this spare time to do extra work or get ahead on study activities.
- If there is a legitimate reason why homework or assignments can't be completed by the deadline, it's best to notify the teacher or ask for an extension well in advance. Teachers tend to be understanding and will try to help students, but missing a deadline without communicating with the teacher will not be received well.
- Homework planning is crucial when juggling many subjects and deadlines.
- Plan homework in terms of **priority** (applying the skills from Chapter 3):
  - Not all homework is equally important
  - Not all homework is equally challenging
  - Not all homework is equally urgent
- Use the Smart Priority Grid to work out the relative importance of tasks and the urgency based on deadlines and prioritise accordingly:
  - As a rule, it's best to give priority to the most important and most challenging tasks first, obviously balancing this with their relative urgency. Sometimes more challenging or time-consuming tasks get pushed to the back of the queue due to avoidance and procrastination, but it's best to tackle these first and get them out the way so they don't hang over you.

  - o Everyone has a different focus rhythm – some focus better in the morning, for others it's after school or in the evening. Try and plan tasks requiring the greatest level of focus at a time when you know you will be able to best concentrate, and leave some of the easier, quicker, and more mundane tasks for times when you might be tired or find it harder to focus.
- Apply the **80/20** principle when possible:
  - o Work out which tasks/parts of tasks will yield the best return on effort and prioritise those.
  - o As a rule – spend more time and effort on tasks which are complex, important, and likely to have a greater impact on future academic outcomes (such as maths problems or English practice paragraphs).
  - o If a task is simple, or less important (e.g. creating a subject title page), be as efficient as possible and don't devote excessive or unnecessary time.
- Understand the point of **diminishing returns** – even for important tasks:
  - o It's unnecessary to strive for perfection on every task
  - o On some tasks it's fine to make it "good enough"
  - o If additional time spent perfecting something is disproportionate to the learning, value added, or the consequence of it not being 100%, it might be ok to leave it and move onto the next priority in order to progress with other tasks (these judgements get easier with experience).
- Mathematics homework is usually high priority because it:
  - o Tends to be assigned every lesson, building up quickly if allowed to accumulate, so falling behind makes it tricky to catch up
  - o Is usually **urgent**, as it's expected to be completed before the next lesson, meaning deadlines are tighter
  - o Is **important** – if homework and learning consolidation (through practice) is not done (especially if the content is a bit challenging), it creates a shaky foundation making it harder to keep up as topics build on prior knowledge and understanding and increase in complexity.

- **Important note**: maths homework is mostly self-marked at high school, because teachers don't have time to mark the volume of homework that students are required to do. This is a very different approach to primary school, and many students initially don't realise that they need to be proactive and own this process. Marking is often forgotten or perceived to be lower priority – it's easy to fall into the trap of assuming that if the allocated questions have been done, the homework is complete and marking is secondary or optional, but this is a dangerous misconception. Marking is a crucial element in maths learning and should be a priority task.
- It is highly recommended to mark the work as soon as it has been done, because:
  - o The recency effect saves time – if you have just completed the questions, marking is quicker, because there is no struggle to remember what you did.
  - o It reveals gaps in conceptual understanding, which can then be addressed.
  - o Students can seek clarity timeously before moving onto new content and compounding misunderstandings.
  - o If left to accumulate, marking will become a more onerous task, and there is a risk that as other work builds up, there won't be time to go back and attend to marking later, meaning crucial gaps might not be resolved.
- Self-marking is not exclusive to maths and is sometimes also expected in languages, sciences, and other subjects. The same principles outlined above would apply.
- **Bonus homework TIP** – highlight (using a colourful highlighter) any complex or difficult homework questions, because these are great to revise before exams.
  - o An alternative would be to keep a separate notebook containing challenging questions to review or retry when preparing for assessments. This is especially useful in maths but can also apply to other subjects.
- Establishing good classwork and homework habits from the start creates a solid platform for academic achievement.

Use a diary

Following on from the previous point – managing and keeping on top of homework – a diary is one of the most valuable planning and organisational tools in your arsenal. I can't stress this enough! Developing good diary habits at the start, will pay dividends later, especially as the workload escalates and assessments and exams start featuring.

Schools may issue a diary at the start of the year, but if your school doesn't, I would strongly recommend buying your own or setting up an electronic diary.

Why use a diary? It…

- Reminds you what you need to do, so things aren't forgotten
- Helps you keep track of deadlines and due dates
- Gives an overview of all commitments
- Is easier to prioritise if you can see everything at a glance
- Allows you to plan and manage your time
- Creates a greater sense of organisation and control.

Smart use of a diary helps you proactively plan your work more effectively and if managed properly, it guides you in the direction of Quadrant 2 in the Smart Priority Grid, which is where you ideally want to be operating most of the time.

Diary management itself is a vital Quadrant 2 activity, being important but not urgent. It is a valuable part of productivity and quality management which only requires a few minutes of attention each day but will provide a big return for minimal effort.

TIP

Take your diary to every single lesson each day, and write all homework that is assigned, along with a deadline.

Try to write homework for each lesson into your diary **as soon as it is given** in class. If you wait until after school, it's easy to forget what was assigned earlier in the day.

An example: It's Tuesday 7th Feb, and you are given the following work in classes:

- Science – complete worksheet by Thursday 9th Feb
- English – read Chapter 1 of the set work by Tuesday 14th Feb
- Maths – complete exercise A and B by Wednesday 8th Feb

In this example, you would write all the work into Tuesday 7th Feb. Additionally, for work with a later deadline (especially if it's not in the same week), you should write the deadline in on the due date as a reminder. Maths and science in this example are both due the same week, so you will likely remember to do them as it's there for you to see throughout the week. English is only due the following week, so if you don't write in the deadline on Tuesday 14th, it may fall off your radar once the page is turned.

> **TIP**
>
> **Once a whole task has been completed tick it off.!** ✓

If a task has not been completed on the intended date, cross it out and transfer it to a NEW date ✗

For example: if you ambitiously intend to complete both maths and science work on Tuesday 7th, but something comes up and you can't complete the science, cross it off Tuesday's list and put it into Wednesday 8th. This will sit in the Wednesday list along with all other work assigned on Wednesday. If you don't 'reschedule' a task, it's easy to forget that it hasn't been done, especially if your diary is busy.

> **TIP**
>
> **If a task has multiple subparts with multiple deadlines – write them all into your diary.**

In the case of partial deadlines or multiple components, it can be easy to lose track of which parts have been completed, what still needs to be done, and when various parts are due, making diary management even more vital.

Lastly – also write in any other commitments you may have, even if they are not related to schoolwork. This helps you to understand your available time and plan accordingly.

Here is a real example taken from a Year 7 student to illustrate a diary in action.

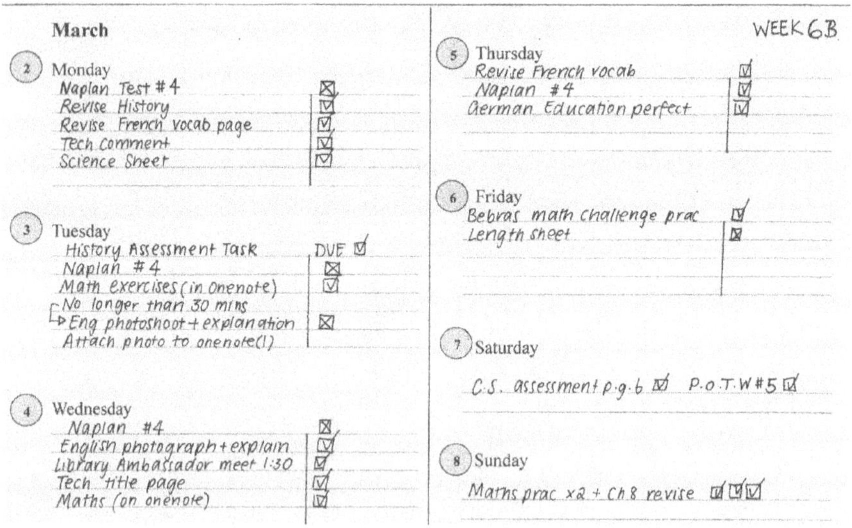

Source: Diary courtesy of I.A.

Figure 5.1 Example of a Year 7 diary

In this Year 7 diary extract, there are a few things to which I would like to draw your attention:

- Each day contains homework assigned for various subjects.
- There are ticks next to work completed and crosses for work not completed.
- You will notice that on Monday everything was completed except "Naplan test #4", so the student put a cross next to it on Monday, and transferred it to Tuesday, then Wednesday, then Thursday, when it was finally completed and ticked off. This kept the task on the student's radar throughout the week.
- In the same way, Tuesday's English photo task was not completed, crossed out, transferred to Wednesday, and ticked when complete.
- A history assessment task was written in for Tuesday, and the student noted this as the DUE date. Using a different colour or capital letters can help things to stand out.
- It's clear that some prioritising took place on Monday:
  - History revision was a top priority because the assessment was the next day. It was therefore classified as important and urgent (Quadrant 1).
  - Subject homework was attended to next.
  - When the student ran out of time, the NAPLAN practice test was transferred. Strategically, this was classified as a lower immediate priority (important but not urgent) since the actual NAPLAN tests were taking place in May – more than two months in the future. This was classified as a Quadrant 2 task.
- A scheduled meeting was written in on Wednesday.
- The weekend contained reminders to do work that was not necessarily homework but preparation for upcoming tasks (Quadrant 2).
- The diary contained tasks that were important and prioritised in order of urgency.

You may wonder why I have spent so much time going into detail on a diary, which is by all accounts common sense and self-explanatory. It's because I have learnt, and seen first hand, that this is often not obvious to students, especially when they are navigating such a big and overwhelming change. Many students believe they will remember everything they have to do, and that spending time managing a diary is a waste of precious time, a useless exercise, or something for "old people". So, why write things down? Well, when things are not written down, organised, and managed, the inevitable result is that deadlines are more likely to be missed, work is done at the last minute or forgotten, and time is spent crisis managing in Quadrant 1, which is very stressful. At the very least, with no snapshot of what needs to be done and what lies ahead, it is very difficult to plan and prioritise, which hinders the ability to work smart.

Note: Students sometimes start off with great diary management, and as they progress further into high school this deteriorates. My advice is to be disciplined with the diary throughout high school. The senior years get very busy, so this becomes an increasingly important tool in organisation and task management.

Effective diary use is a key factor for success, assisting with organisation, time management, and prioritisation, all of which are critical for working smart. If established early on, this will be a great asset in the higher years as the demands grow.

Create a positive and conducive home study environment

Imagine you have done all the "right" things… you have completed your classwork, written all the homework and deadlines in your diary, and you have a plan for what to work on first. You are filled with productive intentions, and then you arrive home…

It's been a long day. You are tired, hungry, and the home environment seems full of temptations which are calling to you. The video games, TV, your phone, social media, YouTube, the refrigerator, your bed. So many alternatives that are more appealing than work. You know you have a lot to do, and you **really** should try to work, so you grab a snack and sit at the kitchen counter armed with mobile phone and laptop. In this scenario the odds are already stacked against you, and the likelihood of being focused and productive is slim. The reality is that to be truly productive, your physical environment is crucial to the effective execution of your intentions. So the aim is to create, as far as possible within the constraints of your individual circumstances, an optimal environment conducive to productive study.

The ideal is a dedicated study space away from noise, distractions, and preferably in a place which isn't associated with sleep or recreational activities. For example, it's not optimal to work in the same space you usually play video games. It's also inadvisable to work in bed. This sends confusing signals to your brain regarding work time vs sleep or relaxation time, so it's not optimal for achieving either good work or good sleep.

If it's not possible to allocate a separate room, don't be discouraged, there are many other ways to enhance a study environment. Some of the key features are:

- A desk or table to work at (even if this is a dining table).
- A comfortable chair, preferably one with good back support.
- Good lighting. If lighting is too dim, supplement with a bright desk lamp. This creates a more conducive study environment and minimises eye strain.
- A space to organise books and files.
- A desktop organiser with drawers is great. They aren't expensive and can be helpful for organising workflow. Ideally you can allocate a drawer to each key subject.

If all else fails, a school or public library is a great place to study and is accessible for most students.

Put devices away… far, far, away!

As an extension of the previous point – creating a conducive study environment – it is vital to implement deliberate strategies to minimise distraction. In simple terms: remove the temptation of technology (unless it is absolutely necessary for your work).

Having a comfortable, organised study environment is advantageous, but the most crucial element is to eliminate distraction as much as possible. Smart phones, smart watches, tablets, and other devices wreak havoc with productivity. They create disruption and temptation for even the most diligent and disciplined students, and the only way to combat this is to limit access. This has always been a non-negotiable in my kids' study spaces. I am sure that parents don't need much convincing on this point, and in my experience it's best to set this boundary early so good habits are developed. It is harder to implement down the track and will invariably be met with resistance. I believe this is a battle worth fighting (along with no devices in bedrooms overnight – but that's a different point).

What this translates to in practical terms:

- Leave the phone, smart watch, tablet and other devices in another room.
- If you don't need to work on a laptop, put the laptop away, and if you have the option to work from a hard-copy textbook rather than an online version, choose the physical textbook every time.
- If you do have to work on a laptop, shut down social media, email and any group chat apps before you begin.

This seemingly mundane issue is of critical importance, and one of the most significant determinants of productivity. If not managed well, it impedes academic performance. This is a big statement, which I believe I need to qualify…

Imagine the realistic and relatable scenario in which a student puts their phone on the side of the desk and opens their laptop to begin their maths homework. The student may have good intentions to shut out distractions and focus on their work. Before long, the phone pings with a couple of text messages relating to the sports fixtures for the next weekend. The student glances at the phone to briefly assess the importance and urgency of the messages, dismisses them and gets back to work. Next, an email pops up from the head of the year group to remind students about school photos the next day. Given the email was from a teacher, the student has a quick look to see what it's about. They then return their attention to the maths work. A few minutes later a message pops up on the Teams chat to negotiate a meeting time for the history assessment group task due the following week. Three of the group members respond immediately with their availability, so the student feels compelled to check their diary and reply promptly with their own availability. Twenty minutes later an Instagram alert pops up with a stream of new messages, which the student quickly checks, and while in Instagram, a recommended reel grabs their attention, which just takes 30 seconds to watch. It's true that some of these alerts were legitimately school related, and may have even been important, but in all cases they could have been checked and responded to later. Of course, the last couple of interruptions were pure distraction, with no importance. This string of disruptions had the following implications for the student (even if they were not aware):

- Multi-tasking meant switching in and out of maths to attend to other tasks, which invariably interrupted and reduced focus, efficiency, and productivity. In addition to the maths taking longer than it needed to, quality and accuracy were most likely also compromised in the process.
- The last couple of interruptions took the student away from Quadrants 1 or 2 and placed them squarely in Quadrants 3 and 4.
- The extra time taken to complete maths had a ripple effect on other homework, so the student was unable to complete everything they had intended for that day.

This scenario is a common one, even for adults, and even the most disciplined students are largely powerless to resist the temptation of checking a digital notification. The upshot of this is that the presence of electronic devices in a study space will inevitably compromise quality and efficiency of study, and over time this accumulates, creating obstacles to academic success. It takes you further away from working smart, and detracts from the benefits of the 80/20 principle, because it often entails working longer (and harder) with a lower return on your time and effort. It has been my observation that students who take control of their technology, rather than letting technology control them, tend to produce better quality work in a more effective manner, which translates to superior academic outcomes.

Having said this, there are ways in which technology, if used appropriately, can be beneficial in supporting learning. Some examples of this are:

- Apps that track your study progress by recording time spent on various tasks or subjects. Some of these apps can even be group based, allowing you to see how much time your peers are studying. This can be positive or negative and is different for individual students. It could be motivating, or it could be stressful. (If it creates stress and pressure, it's best not to use it.)
- Apps that allow you to create and manage to-do lists are very helpful, because they can be used as strategic planning tools.
- Apps that block access to certain sites for designated periods of time, facilitating uninterrupted study, are also helpful for those that may struggle with self-discipline.
- Apps that help organise study notes or create revision flashcards can also be useful.

Be cautious and realistic when assessing which apps are truly valuable – some may appear useful, but in reality they consume time and attention without making a positive contribution to working smart.

Plan when and how to structure home study time … and guilt-free relaxation

It's highly recommended to plan your home study time, also known as "independent learning". This includes time after school on weekdays and weekends, and in later years it will also include school holidays. Having a predetermined plan to structure your time is helpful in keeping you on the right track, especially when you don't feel like working. If there is no committed plan in place, it's too easy to procrastinate, tempted by Quadrant 4 activities, leading to inefficiency and time wasting. You might even be tempted to put work off altogether, leaving it for another day. The danger is that the work piles up and becomes overwhelming to tackle later, leading to more time spent in Quadrant 1 chasing your tail, and experiencing additional stress that could have been avoided. Alternatively, you risk landing in a downward spiral of incomplete work, with no time to catch up, which will be detrimental to academic success. Planning is therefore at the heart of working smart and falls into the most valuable 20%. Time spent planning will fall into either Quadrant 1 or Quadrant 2 in the Smart Priority Grid, depending on how far ahead you are planning. Either way it is essential to success.

The additional benefit of structured work time is that you can also plan dedicated breaks for guilt-free enjoyment. Time to recharge is important and should be planned into your routine. There are various options for structuring study time, and different methods work for different people. I will outline some tried-and-tested approaches and hopefully you will find a method that suits you.

TIP

Create a study planner/timetable for evenings/weekends (and holidays if required).

- This can be done weekly or set up for a month or a term.
  - o In the early years, a weekly or fortnightly plan for day-to-day work should suffice, but later when the scope of work and deadlines are greater, or you are approaching an exam block, a longer-term plan is helpful.
- It can be paper or electronic – Excel spreadsheets are effective and can easily be edited, copied, and saved. They are always accessible and less likely to be lost.
- It's useful to include other commitments you have, so you can work around these:
  - o For example, if you have sport on Wednesday evenings, block off this time as it's not realistically available for you to study
  - o Also include any ad hoc appointments or commitments so you can work around these (e.g. dentist appointment)
  - o Create space for exercise, mealtimes, and travel time.
- Once you have an overview of the week's commitments, you can see what time can be allocated to homework or study, enabling you to plan your study schedule.
- Allocate work timeslots in your planner – 30-minute study increments tend to work well for most people:
  - o 30 minutes suits the minimum attention span of most students
  - o 30 minutes is a psychologically manageable chunk of time, reducing overwhelm and fatigue
  - o 30 minutes can be enough time to complete shorter or chunked tasks
  - o … but importantly can also easily be scaled up to create longer sessions in multiples (60, 90, 120 minutes, or greater timeslots as required)
  - o It allows you to make productive use of small gaps of time in a structured way – e.g. half an hour before dinner, which might otherwise be wasted
  - o It creates an easy system to accommodate different subjects and priorities.

TIP

Be realistic in your plan!

An overly optimistic plan with study slots crammed into every available minute won't be sustainable and will create an unnecessarily pressurised and stressful situation. It's also disheartening if an ambitious plan results in non-completion of tasks, because that work needs to be rescheduled for another time, creating a sense of falling behind, which can be demotivating for even the most diligent student. Try to find a balance between making constructive use of your available time, while also building in flexibility to accommodate for things taking longer than anticipated, or for inevitable curveballs.

TIP

Schedule time in for the process of planning, organising, goal setting, and goal review.

- Although planning is vital, the process takes time, so students may dismiss this as wasteful because it's not a school-assigned task that can be completed and ticked off in the diary. However, good planning, goal setting, and organisation is the opposite of wasteful – it's essential to working smart, saving time later. Even 15 minutes spent prioritising which tasks to focus on, and whether to assign a short or long timeslot, provides crucial direction to maximise productivity in line with your goals.
- It's beneficial to allocate a short amount of time each week or fortnight to review progress on goals or set new goals. Reviewing at regular planned intervals helps you to stay on track and allows you to make timeous adjustments.
- It's also important to set aside time for filing, organising resources, preparing for school the next day, updating your diary, checking school email, etc. These tasks are often given lower priority because there are no due dates or school deliverables, but they

are important Quadrant 2 activities which, if kept under control daily, make it easier to stay on top of your workload. Ideally, filing should be done daily, but if this is not feasible, it should be done weekly. If allowed to accumulate, things get lost and it becomes difficult at exam time to remember what resources you have and where to find them. If done regularly, filing only takes a few minutes, but when left unattended it becomes a more onerous task. This important organisational skill will pay dividends later, as good organisation provides students an academic edge.

Here is an example of how a Year 7–10 independent study/homework timetable may look (Note: Required study time will likely be less than this for Years 7–8 and more for Years 9–10):

	1-7 Feb						
	Monday	Tuesday	Wednesday	Thursday	Friday	Saturday	Sunday
7:00 am						Sport	
7:30 am						Sport	
8:00 am						Sport	
8:30 am						Sport	
9:00 am						Sport	Study
9:30 am						Sport	Study
10:00 am						Sport	
10:30 am							Study
11:00 am			SCHOOL				Study
11:30 am						Study	
12:00 pm						Study	Plan/organise
12:30 pm							
1:00 pm						Lunch	Lunch
1:30 pm						Lunch	Lunch
2:00 pm						Study	
2:30 pm						Study	
3:00 pm							
3:30 pm							
4:00 pm	Travel home	Travel home	Sport	Dentist	School meeting		
4:30 pm	Break	Break	Sport	Travel home	Travel home		
5:00 pm	Plan/organise	Plan/organise	Travel home	Break	Break		
5:30 pm	Study	Study	Break	Plan/organise	Plan/organise		Exercise
6:00 pm	Study	Study	Plan/organise	Study	Exercise		Exercise
6:30 pm	Exercise	Exercise	Study	Exercise			
7:00 pm	Dinner	Dinner	Dinner	Dinner	Dinner	Dinner	Dinner
7:30 pm	Dinner	Dinner	Dinner	Dinner	Dinner	Dinner	Dinner
8:00 pm	Study	Study	Study	Study			
8:30 pm	Study/Prep	Study/Prep	Study/Prep	Study/Prep			
9:00 pm							
9:30 pm							
10:00 pm							

Sport/exercise
Appointment
Study/Plan/Prep

Figure 5.2 Example of a study/homework timetable

You will notice a few things:

- The plan is realistic, allowing for travel time, mealtimes, appointments, and extracurricular or personal commitments.
- It is not crammed full, and there are gaps which can be used to slot in more study time if a task takes longer than expected.
- It is constructed in 30-minute intervals but allows for some time slots to be upscaled to an hour (or more on weekends).
- Planning and organising are formally built in at various points:
    - o At the start of independent learning each day there is time for planning, prioritising and decision-making about how best to allocate work, as well as time for filing or organising resources.
    - o At the end of each weekday and the end of the weekend there is time dedicated to preparing and organising for the next day or the week ahead.
- There is time set aside for exercise.
- Breaks are planned into the timetable.
- Allowance has been made to facilitate at least 8+ hours of sleep.
- Note: During assessment periods the schedule would alter to accommodate additional study and preparation.
- Note: Although the same process and considerations are applicable for a study timetable in higher years, you can expect the schedule will be tighter, with longer days, later bedtimes, and fewer gaps as there will invariably be more commitments, a heavier workload and more study time required, especially in Years 11–12.

Theories on how to best plan study time and optimal session length

- One very popular method of structuring study time is the **Pomodoro® Technique**:
    - o Developed by Francesco Cirillo in the 1980s as a time management tool.

- o The premise is that productivity can be boosted by working with focus for a time interval of around 25 minutes, followed by a short break of 5 minutes.
- o The short time interval facilitates focus, keeping the mind in a state of alert concentration, reducing the propensity for procrastination and distraction.
- o This method claims to create greater control over how time is spent, thereby improving efficiency.
- o The short intermittent breaks can alleviate intensity, reducing mental fatigue.
- o The time intervals can be adapted to suit the task and the individual, making it a flexible tool.
- o For more information, refer to https://www.pomodorotechnique.com.

– It's significant to acknowledge that while 25–30 minute blocks of time, as per the above method, suit some students, it does not work for all students or all tasks. There are occasions when short blocks of time can also be disruptive and create inefficiency. At times a student may naturally find themself fully absorbed in their work and reaching a state of "flow", where they can spend hours focused on one task. An example of this may be a student working on a painting, playing a musical instrument, reading, writing an essay, or working on a complex maths problem. When this happens, it's important to capitalise on this state of focus, and avoid interruption, as it tends to go hand in hand with increased productivity and performance, leading to better results.

– This therefore brings us to another valuable perspective in study planning, which centres on the concept of **flow**. It's important to understand, as it's an ideal state to aim for in the pursuit of excellence. **Flow theory** was first introduced by researcher Mihaly Csikszentmihalyi in the 1970s, and his work, alongside Dr Martin Seligman, was foundational in the development of positive psychology, which is widely used in many schools today. Csikszentmihalyi describes flow as "being in the zone", and "being fully engaged in an activity combined with a state of optimal experience". He also states that "flow helps people achieve what they thought was impossible" (Csikszentmihalyi, 2009, as cited in Sutton, J. 2016). When it comes to planning work schedules, it's therefore important to allow the time and space for flow to be reached. Imagine the example of a student trying to prepare a music composition or write an English essay. At the 30-minute mark, the student may be fully immersed, making excellent progress and following a natural train of thought. If they cease working at the end of 30 minutes and the flow is interrupted, it could create an unnatural break in ideas that may not be easily resumed, even if it's only 5 minutes later. In that instance, it would have been more beneficial to continue in the state of flow until its natural conclusion. The 30 minutes may have extended into 40 minutes or 2 hours, with a greater likelihood of positive results, whereas the premature interruption could impede productivity, efficiency, and performance. Frequent interruptions or shifts between tasks that would benefit from a period of flow, can approximate the pitfalls of multi-tasking, where resuming work after each break requires a brain reorientation which slows processing, potentially impacting productivity and quality.

- o A flow theory side note and **TIP**: Seligman and Csikszentmihalyi (in Sutton, 2016), posit that flow is at its best when engaging in enjoyable and purposeful endeavours. Students will naturally find it easier to achieve flow when working on subjects or assignments that align with their interests and passions. Thus, it can be more challenging to engage with sustained focus on work that is not as interesting or fun. Realistically, most students do not enjoy the prospect of schoolwork and study (even in subjects they find interesting), so this creates a challenge for achieving flow. This is where the 'purpose' component is crucial, because in the absence of 'enjoyment', it creates the guiding force necessary to fully engage. This takes us back to our earlier point about the importance of setting goals and having a clear vision of achievement. With a defined goal and a sense of purpose, this can provide the motivation or self-discipline necessary to engage with the task in a meaningful way.

- From a practical perspective, it's also necessary to consider that most school examinations are longer than 30 minutes. In the early years, exams may be shorter, but in the later years and certainly in Years 11 and 12, examinations are routinely 2–3 hours. It will be important to start practising and training to focus for longer periods of time (eventually up to 3 hours), ensuring that time is built into the study planner to allow for this. Most high-achieving students in Years 11 and 12 would likely study for much longer periods of time to build their stamina.
- When planning a study schedule or independent learning timetable, it's important to balance the benefits of short periods of sustained focus with the benefits of longer work sessions to optimise flow (and to allow space for both). This is where study planning and time management are crucial, because some tasks and assignments lend themselves to short sessions, while others intuitively require an allowance for longer sessions. Deciding how to plan and manage study time will become refined over time and with practice as students get to understand their optimum working patterns and the demands of the various subjects.
- A caution here – while planning is crucial, beware not to fall into the trap of spending all your time planning and re-planning at the cost of getting work done. The plan doesn't need to be perfect – good enough is fine! You need to remember that if the plan doesn't translate to action, it's pointless and counterproductive.

Note: At this point we are only talking about planning WHEN the work can be done – ensuring there is structured time and space dedicated to independent learning. Deciding WHAT work will be done and how to populate your allocated time slots takes us back to the principles of prioritising and time management. This is something we will come back to when we talk about specifics later.

3. Starting as you want to finish: The bigger picture

Make the most of all the school has to offer

There's a lot to navigate in high school with increased academic demands, new routines, skills, and social dynamics. It's important to tap into opportunities and resources whenever possible, as all-round school engagement contributes significantly to success.

- Academic:
  - Many schools provide extra academic resources, such as lunchtime or after-school tutorials with teachers or senior students. These are worth attending for help with homework, assessment preparation, study notes, or even clarification of a new concept learnt in class. This resource is often underutilised because students are unaware, or resist signing up for fear of appearing to need help, but for students willing to engage with these resources, it provides an academic advantage and usually costs nothing. There may be a misconception that these resources only cater for students who are struggling, but it's important to recognise that engaging with extra academic resources equally assists top-achieving students wanting to seek feedback and acquire skills that will take them to the next level.
- Competitions:
  - Many schools provide opportunities to participate in academic competitions which students can sign up or try out for, either as individuals or teams. Some examples might be mathematics or science Olympiads or competitions, ICAS assessments, da Vinci Decathlon, engineering challenge days, literary competitions, etc.

  - o Year 7 students may feel intimidated and reluctant to put themselves forward, but these valuable opportunities provide academic challenge (and sometimes rewards), encourage critical thinking and creativity, build confidence, teamwork skills, and create social connections beyond the classroom. (Also, in our experience, students who miss an opportunity to try out for an academic team, may not get another chance later.)
- Clubs, sport, and music:
  - o Most schools offer a variety of clubs (academic, creative, cultural, environmental, service, etc.) and it's worth signing up for at least one. Co-curricular engagement is viewed favourably and creates opportunities for extracurricular learning, social connection and community contribution.
  - o Involvement in co-curricular activity also contributes to vertical integration within the school, which is very important.
  - o Most schools also offer music/choir/band and sports programmes – for some schools, sport is compulsory. This is another valuable way to engage in activities beyond the classroom, creating further opportunities.
- Leadership:
  - o Some schools offer leadership opportunities from Year 7, while others start in later years. For students with leadership or service aspirations, especially those aiming for a prefect role in Year 12, it's advisable to start contributing early on. Put your hand up for roles available in your year group or house group, as well as those in clubs, sport, and co-curricular activities. Initial attempts may be met with rejection, but a positive attitude, perseverance, and diligence don't go unnoticed. If this is paired with consistent good behaviour and effort, it accumulates and stands you in good stead.

All-round engagement in diverse areas of school life smooths the transition to high school both academically and socially, creating a more enjoyable experience as students become absorbed in the school community, fostering a sense of belonging. It also provides avenues to participate in activities which challenge them intellectually, align with their interests, and connect them with like-minded individuals.

Additionally, these days, schools, universities, and employers place a great deal more emphasis on 'well-rounded' students who possess '21st century' skills. According to Michael Fullan and Geoff Scott in their *New Pedagogies for Deep Learning Whitepaper* (Fullan and Scott, 2014), the most critical and desirable attributes are referred to as the 6Cs – **character, citizenship, collaboration, communication, creativity, and critical thinking**. While schools have successfully incorporated these into the curriculum as far as possible, it is often the complementary co-curricular activities that provide the most valuable opportunity for students to develop, practise, and internalise these skills. Participation will usually be included in school reports and ultimately in the Year 12 transcript, which will be used by tertiary institutions and potential employers when considering applications for endeavours beyond school. This provides a further benefit to engagement in a variety of school and co-curricular activities (within reason, as **too many** activities can create scheduling problems and compromise academic success).

Academic achievement certainly opens the door, but to cross the threshold, students must demonstrate a well-rounded skill set. This is best built steadily from Year 7.

Eat well, sleep well, exercise well, and moderate screentime

The best laid plans will come undone in the absence of adequate sleep, nutrition, exercise, and screentime moderation. The science is clear – optimal performance depends on a healthy body and healthy mind. Research indicates that although diet, sleep, physical activity, and screen management are individually important, it is the cumulative effect of these that enhances academic performance. Starting high school with good behavioural routines, and maintaining this throughout, will increase academic performance and improve mental health.

Although there may be occasions where you have a late night, skip exercise, or indulge in an excess of screentime, making this the exception rather than the rule, and maintaining overall healthy habits, is important in creating conditions conducive to study – helping with focus, cognitive acuity, and the processing and retention of information. Here are some guidelines for secondary school students:

- Aim for 8–10 hours of sleep per night (Mayo Clinic, 2023)
- Aim for 60 minutes of moderate/vigorous exercise per day (can be made up of shorter periods of activity throughout the day) (Australian Government Department of Health, 2023)
- Try to include food from all five food groups in a balanced diet (exercising caution regarding food allergies) (eatforhealth.gov.au)
- Do not exceed two hours per day of sedentary recreational screentime (Australian Government Department of Health and Ageing, 2026).

A couple of final TIPS …

- Always save and back up your work!
  It can be soul destroying to lose hours of work that has not been saved and backed up. There are various ways to ensure this doesn't happen:
  - Check your autosave option is 'on'
  - Actively click 'save' intermittently AND before closing your work
  - Work in an online environment, accessible from any device (e.g. Google Docs)
  - Save to an external hard drive or the cloud
- Always keep your locker locked and NEVER share your locker code ☺

Beyond Year 7...

So far everything in this study guide has centred around starting well and creating systems and processes for successfully navigating the first year of secondary school. The Year 7 platform is the foundation on which the remaining five years of high school is built. The same skills, strategies, and techniques outlined in the first five chapters apply throughout secondary school, although they will be adapted, elevated, and finetuned as the academic demands grow. If good habits are well entrenched during Year 7, these tools become second nature, and well-practised skills become intrinsic to how you approach the challenges that will invariably arise, placing you in the best position to succeed. Mastering the skills and techniques outlined thus far reduces the overall cognitive load, freeing up valuable mental space and time required to achieve excellent results.

PART 2:

HOW TO MAXIMISE YOUR GRADES

"Having the knowledge is one thing, but putting it into practice is another" (A.A. – 99.5 ATAR)

CHAPTER 6:
Overview of curriculum, subjects, and assessments

We have spent significant time discussing general study advice and establishing processes, frameworks, and organisational techniques to lay a solid foundation to achieve your best. BUT to take things to the next level we must get into the detail of exactly HOW to strategically approach individual subjects and successfully navigate the complex assessment system to optimise performance.

While the assessment programme is only one facet of high school, conquering it is arguably the key to academic achievement, as this is where marks are earned and lost.

It is where grades are determined and ATARs are forged.

In this chapter, I will share some valuable and pragmatic strategies to facilitate working smart within the assessment domain. Some will be applicable to all subjects, while others are subject specific.

Note: Most of the information and strategies in the following chapters will be applicable to all year groups including Year 12. Because high school is a process of cumulative skill building, almost everything is taught and approached with the end in mind. Skills learnt in Year 7 become increasingly important in Year 12, so the skills, techniques, and tips outlined here are relevant whether you are a Year 7 or a Year 12 student. Where requirements are specific to a stage, I will point this out.

Before delving into this, I think it would be useful to provide a brief orientation to the subjects in the NSW curriculum, and the school streaming and assessment processes …

1. Curriculum and subjects

In most schools, Year 7 students undertake an array of courses to provide the opportunity for exposure to different areas of potential interest across STEM, humanities, languages, physical education, music, creative arts, etc. The core NSW curriculum in Year 7 is centred around English, mathematics, science, history, geography, PDHPE (personal development, health and physical education), languages, technology, art, music, and sometimes religious studies. This volume of diverse subjects can create a significant workload, and **ALL subjects are examinable, cumulatively contributing to overall performance.** It's crucial to understand the requirements and workload of each subject, and to plan, prioritise, and manage time well to successfully juggle up to eleven subjects simultaneously, especially during assessment time.

As students progress through high school they can tailor their courses and the level at which they study, in line with their strengths and interests. This means the number of subjects decreases over time, but the workload associated with each will increase. We will go into detail on subject options, choosing electives and HSC courses later.

The most important thing to note from the outset is that currently English is the ONLY COMPULSORY subject for the HSC.

2. Academic streaming

Sometimes pre-intake assessments for numeracy, literacy, and general aptitude are conducted prior to Year 7 to enable academic streaming, particularly for English, maths, and science. This means that in many schools, students of similar ability will be grouped into classes to permit targeted teaching. This grouping is usually not cast in stone and can change depending on student performance once the Year 7 assessment programme gets underway. It's not uncommon for students to be initially misallocated, as a single streaming assessment in Year 6 does not always provide an accurate representation of student ability. Once school starts there can be movement, with students potentially shifting into higher or lower streams as appropriate. This streaming process can be quite fluid throughout the first few years of high school as the groupings may be reviewed at the end of each academic year based on student results.

Generally, top streamed classes move at a slightly faster pace and may explore some topics in greater depth to accommodate the extension needs of students. This can be subject and teacher specific. Having said this, in the formative years most classes within a school will study the same curriculum and set works, and sit the same exams for most subjects, irrespective of stream. It's usually in the later years that content is formally differentiated, often determined by the level at which students choose to study certain subjects.

Mathematics can be an exception. At some schools, the top streamed class(es) may sit differentiated or additional exams from the start (usually accompanied by a heavier workload), although it's still possible to move up and down within the streaming structure each year (until a certain point). Select schools offer formal acceleration programmes beyond regular streaming, and I will touch on this later.

Generally subjects such as PDHPE, art, technology, music, languages, and religious studies are not streamed, and the cohort is mixed ability to extend social engagement within the broader year group.

It's important to note that while streaming can be helpful for schools to align teaching to student needs as best they can, the stream by no means defines student performance…

The power to maximise academic potential and achieve stronger results lies within each individual student and is determined by their effort and ability to apply the skills and strategies for success.

3. Assessments and exams

Formal assessments and exams tend to dominate secondary school … at least this is the way it feels for most students. It's one of the biggest changes from primary school and it's the aspect likely to create the highest degree of stress and pressure. The start of Year 7, and the onslaught of assessments and exams can be overwhelming, and it's easy to feel lost in the system. In fact, this is equally true for every year. Being forewarned and mentally prepared is helpful, because students who start on the front foot, managing the process effectively from the beginning, tend to build greater confidence, embed key skills, and set themselves up to excel.

In the following chapters, I will share practical tips and insights (and a few explanations) that I personally believe would have been beneficial for us to know from the start. Some of these may seem intuitive, and many of them would likely be revealed sooner or later, but there is advantage in knowing and understanding from day one, because the sooner you know, the sooner you can apply them. Some things we worked out by chance, some by hard-learned experience (which we would have preferred to avoid), and some only became clear much later. Despite the best intentions of highly skilled educators and administrators, there is a fair amount of "assumed knowledge", which I wish had been made explicit. Teachers focus primarily on teaching subject content (which is, of course, crucially important, and their main role), but in our experience not much time is devoted to teaching the specific skills of **how** to apply content knowledge to maximise success. Schools may run student workshops on study skills, and good teachers provide valuable feedback on **what** to do better, but students are often left to figure out on their own **HOW** to implement these ideals. My goal is therefore to provide key information early, which will hopefully smooth the journey and create optimal conditions for academic success.

Schools have some autonomy regarding how and when they assess students, but the overall process is governed by NESA. It's worth taking time to unpack and understand the assessment system because assessments pose a great challenge for many students, but for those who can work smart, approaching assessments in a tactical and strategic way, it becomes the key to unlocking academic excellence. With this in mind, let's examine some key aspects of assessment.

<u>Number, frequency and timing of assessments/exams</u>

- Number of assessments can differ by school, but each subject is usually assessed once per term, meaning that a Year 7 student can expect approximately 30–40 assessments per year.
- Some schools don't examine each subject every term in Years 7 and 8, which can result in fewer assessments.
- Conversely, some schools split assessments into multiple smaller components, which entails a greater number of assessment tasks overall.
- Some schools spread assessments through the term, meaning students can undertake multiple assessments spanning over many weeks.
- Some schools schedule dedicated assessment blocks where all assessments are condensed into a week or two each term.
- In the more senior years, the volume of subjects drops, with a corresponding reduction in the number of assessments.
- Senior exams, especially those at the end of the academic year, and certainly for Year 12, tend to take place in scheduled exam blocks.
- In Year 11, you usually have three assessments per subject, which equates to one assessment per subject in each term, since Year 11 is only three terms.

- In Year 12, which is four terms, you usually have a total of four assessments per subject, with the 4th assessment being the trial exam, mimicking the HSC final exam in both content examined and length of the exam.

Types of assessments and exams

- Exams are often referred to as assessments (which sounds less ominous), but to clarify, most exams are assessments, and not all assessments are exams.
- Schools have some latitude in how they examine aspects of the curriculum and, as such, there are different types of assessments:
  - Some tasks are **hand-in**, meaning they are worked on/written outside school and submitted in hard copy, electronically, or both (note that these tasks are usually run through a plagiarism checker to identify anything that has been copied or AI generated)
    - Examples of hand-in tasks could be a creative writing assignment, a research project, a practical report, or a video submission.
  - Some tasks are **presentation** based – often these tasks are prepared outside school and presented in class
    - Examples of presentation tasks could be a prepared speech, a PowerPoint slideshow, or a multi-modal presentation.
  - Some tasks are **practical** and performed at school
    - Examples of practicals could be experiments or performances.
  - Some tasks are **group** based and some are **individual**.
  - Some tasks are traditional style **tests or exams** – conducted in class under timed conditions, and these may be written or in the form of an online quiz.
  - Tasks can be **formative or summative** or have elements of both.
  - Tasks can have **multiple components** with more than one submission requirement or deadline

        - For example: a research project may have a report accompanied by a process log, or a hand-in composition may be followed by an in-class reflection.
- Some tasks/exams cover **select content**, while others cover all **yearly content**
    - In Years 7–8, assessments/exams often test smaller sections of content, with the curriculum examined in sequential chunks, often only testing one term's worth of work, even at the end of the year
        - E.g. science might test: Task 1 – ecosystems, Task 2 – properties of matter, Task 3 – forces and Task 4 – body systems.
    - In later years (sometimes from Year 10 and certainly in Years 11–12), the final exams (also known as yearlies) are cumulative, examining all content learnt during the year, including topics or modules that have already been progressively tested throughout the year.
    - Maths is a notable exception, since many schools, from Year 7, conduct yearly exams testing all content from that year.
- Schools are required to create a mix of assessment types, giving students the opportunity to showcase their learning and simultaneously develop broad skills in written and verbal communication, multi-media, presentation, practical application, research, collaboration, critical thinking, and problem-solving.

Assessment requirements

There are certain requirements and protocols for school assessments, along with strict guidelines, which are particularly rigorous for Years 11–12. Schools will provide students with comprehensive assessment policies. These policies will, among other things, outline the process, rules, and expectations around student conduct. They also clarify consequences of student malpractice or insufficient assessment participation. It's worth reviewing these rules carefully so students (and parents) have a clear understanding of their rights and responsibilities in relation to school assessments.

Assessment notification

Assessment notifications are documents with which you will become very familiar. These are extremely important, as they provide all the information and instructions for each assessment or exam. An assessment notification is mandatory and will be formally issued in writing for each assessment task at least **fourteen calendar days** prior to the task, ensuring students are given adequate time to prepare. Sometimes, depending on the nature of the task and school discretion, students may be given more than two weeks' notice. Students should be given a hard copy of the notification or alerted if it's posted online. It's worth spending a bit of time on this, because proper attention to the assessment requirements in the notification is a key determinant of success.

As a rule, the following information (if applicable to the task) will be included in an assessment notification:

- The **nature of the task**, for example, written or practical exam, presentation, etc.
- The **due date and time** for submission, or the date and time of the exam
  - o Late submission (even if it's only a few minutes) could carry a penalty, such as deduction of marks
  - o If the task has multiple parts, the due date for each component is specified.
- The **length** of the exam (if applicable), e.g. 80 mins + 10 mins reading time.
- Relevant **task limits**, such as word limit or presentation time limit
  - o Going over the time or word limit usually carries a penalty – there may be a cut-off point beyond which the work is not marked.

- The number of **marks** awarded in the task, for example, 35 marks.
- The **course components** and component **weighting** – each assessment task will be given a weighting. Not all assessments during the year will carry equal weight in their contribution to the final grade, but the total weights for all assessments for that subject over the year will add to 100%. For example, in maths the tasks may have the following weights:
  - o Task 1 = 15%, Task 2 = 25%, Task 3 = 25%, Task 4 = 35%.
  - o By outlining the assessment weight, students know how much a particular task contributes to their final year mark. This is important as it helps with planning and prioritising – for instance, a task with a weighting of 30% will command greater priority than a task counting 5%.
  - o Additionally, there can be component weightings within an assessment task, for example, a task with a weighing of 15% might have two components with Part 1 weighted 10% and Part 2 weighted 5%. This is also important for students to know so they can plan and prioritise accordingly.
- **Outcomes** being assessed – this relates to the prescribed syllabus.
- Comprehensive **instructions and task details** – the task requirements are explicitly outlined so students know exactly what they have to do.
- **Submission** instructions.
- **Equipment** list – for example:
  - o Maths exam equipment will usually include pens, pencils, eraser, ruler, compass, protractor, and NESA-approved calculator
  - o Geography exam equipment often includes pens, pencils, eraser, ruler, calculator, and string ☺.
- Information regarding student **absence** and protocols for illness and misadventure.
- There may be **topic checklists, textbook references,** or information on **syllabus dot points**, providing guidance to students on exactly what content to study.
- There may be suggested **strategies for preparation**, including practice exercises, past papers, or formative submissions (which

are marked for feedback but not included in year marks). Additional study or revision resources may be provided.

- There will be **marking criteria**, also sometimes called **mark schemes** or **rubrics** (if appropriate for the type of task) – this provides detail on what the marker is going to look for when awarding marks. It outlines the specific criteria that must be met to earn certain marks and grades.
  - For example, if an essay is out of twenty marks, the marking criteria descriptors will indicate which specific elements students should include or explicitly address to earn certain marks/grades.
  - The marks and grades may be grouped and represented as follows:
    17–20 (A), 13–16 (B), 9–12 (C), 5–8 (D), 1–4 (E).
  - The marking criteria page is often placed at the end of the assessment notification and is one of the most crucial pieces of information. As such, I will expand further on this shortly!
  - Note: Marking criteria are not usually issued in advance for maths, science or other quantitative assessments because these questions are not assessed against qualitative schemes, rather they are objectively correct or incorrect.
  - Marking criteria are also not usually included in the notification for in-class written examinations as each question has its own set of marking guidelines. In this case, marking guidelines are often provided retrospectively, enabling students to review and reflect on their answers in relation to the guidelines.

Some examples of what *assessment notifications may look like

*Note that these examples are fictitious and have been created with the express purpose of illustrating what assessment notifications **may** look like – nevertheless, they are an appropriate representation based on the hundreds of assessment notifications I have reviewed from multiple schools over the past ten years.

Year 9 English – Summative Task – Assessment task 1

Task: Creative writing: short story about "a significant moment in life" This is a hand-in task

Date: Monday 23 March, week 9A, Term 1, 2026
Due by 8.20am

Length: Word limit of 850 – 1000 words

Marks: 20 marks

Weight: 20%

Outcomes: *(This will contain prescribed syllabus outcomes)

This notification was distributed to students on 9 March 2026

Task details:
Interview a family member, friend or member of your community who has experienced a significant moment in their life. Use this as inspiration to create your own fictional short story about *'a significant moment in life'*.

Task preparation:
- Conduct the interview prior to week 8
- Complete the writing practice tasks given in class
- Submit a 200-word excerpt of your draft story to your teacher for feedback by 18/3/26
- Implement feedback provided by your teacher
- Review your class notes on creative writing features

Submission details:
- Use the template provided to you (font size 12, Times New Roman, double spaced)
- Save your file as: surname_firstname_Englishtask1_2026
- Submit your task via the Dropbox link before 8.20am on 23/3/26
- Submit a printed copy to your teacher in your lesson on 23/3/26

For illness or misadventure: please refer to the assessment policy document.

For any questions, please discuss with your teacher.

For the marking criteria, please refer to the next page.

Figure 6.1 English assessment task notification

Marking criteria	**Mark/Grade**
• Demonstrates a sophisticated ability to compose an original short story with an authentic and engaging character, setting and complication with regard to a significant moment in life • Demonstrates a sophisticated portrayal of context and values of the character • Displays sophisticated control of language appropriate to audience, purpose and form	A 17-20
• Demonstrates a substantial ability to compose an original short story with an authentic and engaging character, setting and complication with regard to a significant moment in life • Demonstrates a substantial portrayal of context and values of the character • Displays substantial control of language appropriate to audience, purpose and form	B 13-16
• Demonstrates a sound ability to compose an original short story with an authentic and engaging character, setting and complication with regard to a significant moment in life • Demonstrates a sound portrayal of context and values of the character • Displays sound control of language appropriate to audience, purpose and form	C 9-12
• Demonstrates a limited ability to compose an original short story with an authentic and engaging character, setting and complication with regard to a significant moment in life • Demonstrates a limited portrayal of context and values of the character • Displays limited control of language appropriate to audience, purpose and form	D 5-8
• Demonstrates a basic ability to compose an original short story with an authentic and engaging character, setting and complication with regard to a significant moment in life • Demonstrates a basic portrayal of context and values of the character • Displays basic control of language appropriate to audience, purpose and form	E 1-4

Figure 6.2 Marking criteria

Note: I will refer to this marking criteria example later in more depth.

Year 8 Mathematics – Assessment task 1, 2025

Date: Thursday 20 March 2025
Time: Period 2
Length: 70 minutes
Marks: 45 marks (10 marks multiple choice + 35 marks short answer)
Weight: 25%

This notification was distributed to students on 6 March 2025

Task – in-class written exam:
The topics to be assessed are:

1. Algebra
    - Mathematics textbook Chapter 1
2. Equations
    - Mathematics textbook Chapter 2
3. Measurement
    - Mathematics textbook Chapter 3 (A-E)

Equipment list and additional information:

- Students will write their answers into the question paper provided
- Approved calculators are permitted in this task
- Students must bring: geometry set, ruler, pencil, eraser, pens, sharpener, scientific calculator
- NO reference materials, notes, or textbooks will be allowed

Suggested strategies for task preparation:

- Write a brief set of notes on each topic
- Complete the chapter review sections in the textbook
- Complete the assessment past papers provided
- Complete the revision exercises and materials provided
- Review topic tests completed in class

Outcomes: *(This will contain a list of specific syllabus outcomes)

For illness or misadventure: please refer to the assessment policy document.

For any questions, please discuss with your teacher

Figure 6.3 Maths assessment task notification

Note: The structure of assessment notifications, task requirements, and marking criteria will vary slightly for different schools, subjects or task types. For example, a science notification may include detailed checklists of content to study, or a geography presentation notification might include a scaffold of questions to be answered.

<ins>**TIP**</ins>

Go through the assessment notification thoroughly and more than once:

- **In class**
- **Independently**
- **And again before completion, submission, or exam!**

Teachers usually explain the task requirements in class, giving students an opportunity to ask questions or seek clarification. It's crucial to pay attention and make a note of explanations or instructions further to what is printed on the document. Listen to questions your peers may raise and ask questions of your own.

Hint: Teachers often provide their own insights and guidance to assist students in their approach. This priceless information can go beyond the notification and is golden.

The key to performing well in any assessment or exam starts with fully understanding what is required of you.

Any confusion that remains unresolved will at best consume more time later, creating inefficiency, unnecessary time pressure and stress, and at worst lead to missing the brief. This could potentially land you in the unfortunate position of working incredibly hard and investing many hours, only to eventuate in a poor mark – a classic example of working hard without working smart. Either way it will ultimately hinder your performance and compromise results.

Time spent going through the notification in detail is essential for success and constitutes part of the most valuable 20% of effort. Depending on the lead time, this would fall into either Quadrant 1 or 2 in the Smart Priority Grid. It forms the basis for all work that follows and is high priority for working smart.

***

Now you know a bit more about the assessment process and have some idea what to expect when you are issued an assessment notification… but this is just the beginning…

In the next chapter, we will take a closer look at **WHAT** to do next, and **HOW** to strategically approach assessments and exams to create conditions for academic success.

CHAPTER 7:
How to conquer assessments and exams

Before going into the detail and logistics of assessment planning (which I will address later), I believe there is great value in first learning **how** to approach assessments and exams strategically. This involves unpacking what the overall tasks or specific assessment questions are asking you to do and developing skills to meet the brief.

Regardless of the nature of the task, the subject, or the year group, there is an **overarching set of techniques** which can be learnt, equipping you to provide **exactly what is required to ace your assessments**. These techniques can be applied equally at a broad assessment level and to individual exam questions. Following the simple guidelines in this next section will ensure that you:

- Will always know what **type** and **level** of information to provide
- Will always provide **enough** – so you avoid falling short
- Won't provide **too much** – so you know when to stop to avoid wasting time or energy on work that doesn't generate extra marks
- Will provide the information in a way that **scores marks**

It's entirely feasible to aim for 100% in assessments (even in Year 12) ... but this requires strategic application. In this next section, I will highlight **how** and where to focus attention and effort, helping you work smart to secure maximum marks.

WARNING....

Before embarking on this next part, I should warn you that it will be information dense and a lot to digest. On first reading, it may seem intimidating and may be hard to contextualise much of what I talk about if you haven't yet been confronted with it. My intention is to plant the seed which will spark recognition as things arise.

Rather than reading this once-off, I recommend you treat it as a valuable reference resource, which can be revisited for guidance as needed throughout secondary school, because the skills are applicable to all year groups and become increasingly important as you progress to higher years.

Additional notes for parents:

Although you are not personally doing the assessments, it gives your child an advantage if you also understand how to navigate the system and decode what's being asked and how to give the marker what they want. I was asked many times by my kids to give feedback on their work (all the way from Year 7 to HSC), and I was only able to do this in a meaningful and effective way once I developed a good grasp of what was being asked and what the markers would be looking for. In the absence of this knowledge, any feedback would at best be superficial and, at worst, misleading.

Many parents will probably read this and believe they are unable to provide helpful feedback because they aren't teachers, or because they finished school long ago, or because they are unfamiliar with the curriculum requirements, the subject content, or the set works – **BUT** I assure you that even in the absence of all that specific knowledge you will still be able to add tremendous value, because the skills and techniques outlined here do **not** rely on specific subject knowledge. I am testament to that; I finished school in 1989, I have not read most of the English set works, I am not familiar with the specifics of the subject content for many of the subjects, and we have already established that I am not a teacher. HOWEVER, I was almost always able to identify **what was being asked**, whether the **question had been adequately answered**, whether the **level of the response was appropriate**, and whether the **marking criteria had been met**. Armed with this knowledge and understanding, I was able to provide feedback which helped identify gaps and areas for improvement AND, more specifically, provided guidance on exactly HOW to address these in a way which translated to higher marks.

Much of this next section is independent of subject content and instead focuses on higher order skills and techniques which are teachable and broadly applicable across year groups and subjects for most assessment and exam questions.

With that said, let's dive in …

<u>TWO vital KEYS to ASSESSMENT SUCCESS</u> 🔑🔑

KEY 1: Understanding exactly **WHAT** is being **asked** and tested

AND

KEY 2: Knowing **HOW** to give the **marker** exactly what they want

The first part is understanding **WHAT you have to know and do**, and the second part is understanding **HOW to do it in the way the marker expects**, so you can score **maximum marks**.

Note: While these are two distinct aspects, they are closely linked with significant overlap. In the section that follows, some points speak directly to one or other **key**, but many of them have elements that assist with achieving both. I will try to signpost the **keys** along the way.

As with anything in life, if you aren't clear on what's required, it's hard to deliver.

Imagine trying to play a game of soccer or badminton when nobody has explained the rules. It would be virtually impossible to play well, or indeed to win, if you had no idea how to play, where to be, or how to score. School assessments are a bit like that, because students are initially thrown in the deep end, often without a roadmap or clear understanding of the rules. They will receive an assessment/exam notification (or a few simultaneously) soon after starting high school. Certain aspects will be self-explanatory, and others may seem superficially clear, but the nuances of exactly what will lead to scoring top marks can remain elusive, and students who forge ahead without properly understanding the terms of engagement will struggle to access higher marks.

With this in mind, let's look at some vital strategies for **decoding what is being asked** and some tools to **give the markers exactly what they are looking for**.

1. WHAT are you being asked to DO?

The logical place to start with any assessment is getting to grips with the nature and scope of the task at a macro level, understanding what you need to **DO**.

You should clearly understand from the assessment notification whether it is:

- A hand-in/submission or in-class exam
- Practical or theoretical
- Written or presentation/speech based
- A group or individual task
- A single deliverable or multiple components

Understanding the nature and structure of the task provides insight into the tangible skills being tested. For instance, are you being tested on your ability to write an analytical essay, or on your presentation skills? Are you being assessed on your creativity, or your application of scientific theory?

Different **types** of tasks require different skill sets, and consequently a varied approach to study and preparation. I won't go into too much detail now as some of this will be addressed later when I outline subject-specific preparation and study techniques, but these are some of the things you might need to **DO**:

- For an in-class written theory exam you would need to:
    - Do notes, revision, practice papers, manage exam conditions and timing.
- For an in-class essay exam you may need to:
    - Do notes, read and annotate texts/sources, gather evidence, revise, memorise, and do timed practice essays.
- For a hand-in, written task you may need to:
    - Research and gather material, draft, edit, and perhaps create a process log or bibliography.
- For a video presentation submission, you would need to:
    - Draft and edit a script, create and edit a video, demonstrate creativity.
- For a group submission task, you would need to:
    - Form the group, assign roles, allocate work, put it all together, and display collaboration skills.

So, figuring out the nature of the task is the first vital step in knowing what you will have to DO. This fits mostly into **Key 1**

NOTE: We have noticed a shift in recent years as schools seem to be reducing hand-in tasks and replacing them with in-class tasks. This is especially prevalent in the senior grades. One of the reasons is the rise of AI, which complicates school assessments. Although there are ways of detecting AI-generated work, it's a bit of a minefield.

One such example in our personal experience was a move away from hand-in practical reports for chemistry and biology, in favour of these being done in class under exam conditions. It's more challenging and stressful for students, but is a more equitable assessment system carrying greater academic integrity.

2. WHAT do you need to KNOW?

The next critical part of any assessment is knowing what content is being tested. This will be outlined on the notification and should be self-explanatory, as it will usually be related to the topic and content studied in class. It may be a prescribed text in English, an event in history, or a selection of topics from a science or maths textbook.

> **TIP**
>
> **Don't make assumptions!**
>
> **Pay attention to the notification details of what content is being assessed.**

Examinable content doesn't always align exactly with classwork. Sometimes only parts of a unit of study are assessed instead of the whole topic. For example, in science it may be part of a module, or in maths it may be certain units instead of a whole chapter.

For yearly exams, quite often (and almost always in the senior years) the entire year's worth of content is examined, even if it has been previously assessed.

So, it's important to know exactly what the assessment covers so you don't waste valuable time studying content that isn't included, while conversely, you don't want to accidentally overlook something that is being examined. This is essential for working smart and fits mostly into **Key 1**

3. WHAT is the QUESTION asking, and HOW do you answer?

Most assessments contain questions or instructions that need to be answered …

For simplicity, I'll refer to them collectively as 'questions' as they all require a response.

The question presentation depends on the subject and type of assessment.

For example:

- There may be a **single question** explicitly asked in the **assessment notification**
    - This might be the case for a hand-in essay task or a research project.
- There may be **multiple questions** explicitly asked in an **assessment notification**
    - This might be the case for an information report.
- There may be **no questions** in the assessment notification itself
    - This would usually be the case for an in-class written exam where the questions (previously unseen) are contained within the exam paper.

The most important thing to note is that regardless of the nature of the task, or the subject or year group, the questions mostly adhere to a pattern, using specific terms, which must be answered in a specific way.

This is such a VITAL point that I will reiterate it to be certain that you don't miss it!

REGARDLESS OF THE NATURE OF THE TASK, OR THE SUBJECT OR YEAR GROUP, THE **QUESTIONS** MOSTLY **ADHERE TO A PATTERN**, USING **SPECIFIC TERMS**, WHICH MUST BE **ANSWERED IN A SPECIFIC WAY**!

This has elements of both Key 1 and Key 2, as answering questions effectively requires you to understand both **what** is being asked and **how** you are expected to answer. 🔑

This is arguably one of the most critical elements in academic success, because answering the question correctly is a crucial determinant of marks and grades.

To put it simply – if you answer the question in the proper way, you will gain marks, and if you don't, you will lose marks.

TIP 💡

Get familiar with the terminology FAST!

A fundamental aspect of unpacking the requirements of any assessment question and understanding exactly what is being asked, lies in deciphering the terminology used, and comprehending **what level of response** is sought, to enable you to give the marker what they are looking for.

From very early on, probably from the first week, you will notice certain 'key words' appearing in homework, assessments, and frequently referenced by teachers in class. These are words such as: outline, describe, explain, analyse, compare, assess, and a variety of others. On the surface, these words seem innocuous and easy to understand, right? They aren't particularly sophisticated or mysterious words, and most students would be using them comfortably in everyday discourse. In fact, we sometimes use them interchangeably when we speak. HOWEVER – these particular words hold tremendous significance and can be tricky to unpack in the school context, because although they appear deceptively simple, each of them requires something **very specific** from students in terms of **how** they answer questions.

For example: the words 'outline', 'describe', and 'explain', which many people might define similarly, actually require completely different levels of response. To put it another way, implicit in these seemingly similar words is a subtle, but crucial, request for completely different types of answers, varying in volume, depth, and analysis of content. For those who are unaware, this can unfortunately create significant obstacles to success. It has been surprising, and somewhat shocking, to discover that there are some students who by Year 11 have still not grasped this or had it adequately explained to them. I believe this is one of the things that falls into the category of assumed knowledge I mentioned earlier, where schools utilise the terms and language early on, presuming that students automatically grasp the nuances. This misapprehension means in some cases students may not be explicitly taught **how to deconstruct and respond to the terms**, creating a situation where there's a disconnect between what teachers are communicating and what students are understanding. This disconnect can easily go unnoticed, creating academic quicksand for students, as success in assessments at all stages and in all subjects depends heavily on a sophisticated grasp of these terms and how to respond to them. I firmly believe the ability to navigate these key words successfully will be make or break academically.

At this point you may be wondering what the key words are and how to figure out what each word requires, and fortunately the answer is very straightforward. NESA has published a 'Glossary of key words' on their website which defines each term. I highly recommend you visit their website to explore the full list with all their detailed definitions. For ease of reference, I have inserted the link here: https://www.nsw.gov.au/education-and-training/nesa/hsc/student-guide/glossary

For our purposes, the words/terms you need to know are as follows (with the most frequently used terms **bolded**):

Account, **analyse**, apply, appreciate, **assess**, **calculate**, clarify, classify, **compare**, construct, **contrast**, critically analyse, deduce, **define**, **demonstrate**, **describe**, **discuss**, distinguish, **evaluate**, examine, **explain**, extract, extrapolate, **identify**, interpret, **investigate**, **justify**, **outline**, **predict**, propose, **recall**, recommend, recount, summarise, **synthesise**.

Before unpacking the approach to these terms, I'll provide my brief and simple interpretation of what some of the common terms mean, and what they are asking for:

Account (for) – provide reasons for something
Account (of) – relay information about an event
Analyse – understand and communicate how individual parts relate to each other to create meaning
Assess – judge the value, quality, results, or extent of something
Calculate – work something out based on given information
Compare – show similarities **and** differences between things
Contrast – show differences between things
Define – provide the fundamental meaning (definition) of something
Demonstrate – show (by using an example or evidence)
Describe – give a description of the main characteristics and features
Discuss – address a point and give arguments for and/or against
Evaluate – this is similar to 'assess', but the judgement or determination of value is made against specified criteria
Explain – link cause and effect; clarify the relationship between things; provide reasons for why and/or how something occurs
Identify – this requires something to be known and named
Investigate – inquiring into something to understand it
Justify – provide reasons or evidence in support of a conclusion or argument
Outline – provide a general overview
Predict – anticipate what might happen based on the information at hand
Recall – remember and communicate what has been previously learnt
Synthesise – integrate information from various places to create a cohesive whole

Important notes and <u>TIPS</u> on the key words/terms:

- These **key words** are sometimes referred to as: **directive terms**, **directive verbs** or **command words**. From this point on I will refer to them as **'directive terms'**, because this sums up their purpose nicely.
- Tune your ear to listen for them at school and look out for them in all aspects of written work, because they provide direction for you in a multitude of ways.
- While the list is long and overwhelming, some are used more frequently than others. I have bolded those that are commonly used.
- There is an implicit hierarchy in these directive terms – some are considered 'lower order' terms and simpler to address (such as identify), while others are 'higher order' terms involving greater complexity (such as explain, analyse, compare). The requirements for various terms are therefore very different in both the length and depth of responses they command.
- Some directive terms simply require a one- or two-word answer
    - E.g. **Identify** might literally involve providing the name of something.
- Some directive terms will involve a sentence or two
    - E.g. **Define** usually involves a definition with essential components.
- Some terms might involve a paragraph
    - E.g. **Describe** or **outline** require more detail including key characteristics or features of something (often requiring content-specific terminology).
- Some terms are more complex, requiring greater insight or application, and involve an extended response
    - E.g. **Explain, analyse,** or **compare** require more in-depth answers that could range between a paragraph and a full essay depending on the nature of the task and the subject.
- Some terms do not require a qualitative response at all
    - E.g. **Calculate** usually requires quantitative problem-solving.

- These terms are used throughout high school from the start of Year 7 until the end of Year 12, and feature in all subjects, although some terms feature to a greater extent in certain subjects, for example:
  - o **Explain** is asked very frequently across many of the subjects
  - o **Identify, define, investigate, predict, describe, compare**, and **justify** appear quite frequently in the sciences
  - o **Analyse, compare, contrast, discuss, assess, evaluate** and **synthesise** are used frequently in English and humanities
  - o **Account, assess, evaluate**, and **investigate** are used a lot in history
  - o **Calculate** and **deduce** lend themselves to subjects like maths and science
  - o In reality most of these terms could be applied to any subject.
- Some of these directive terms are introduced early, and others might only be encountered later in high school as the work becomes more complex. For example:
  - o **Outline, describe, explain, analyse, calculate** would be used from Year 7
  - o **Evaluate** and **synthesise** may only start appearing later and will feature prominently in Years 11–12.

For many students, the most significant and challenging aspect of responding to the directive terms is learning and mastering the technique for answering the **EXACT TERMS** of the question. The clues to this lie in the **subtext** or definition of each term. It is vital that students look beyond their usual interpretation of this common vocabulary and pay close attention to the **implicit meaning** of each. This is such a crucial component in mastering assessments, that it's worth taking the time to delve a bit deeper into the terms. The aim is that by the end of this section, students will have a clear idea of **how** to conquer these terms in a way that translates directly to higher marks.

Note: I won't elaborate on every directive term as some are self-explanatory, and if students pay attention to the subtext, they should know how to address many of the terms intuitively. However, it's worth unpacking some of the less straightforward ones.

Let's take a closer look at some of the tricky, more nuanced directive terms and I will point out, by example, some of the common pitfalls and simple techniques you can apply to ensure your response fits the brief …

Techniques for responding *exactly* to the directive terms

Explain:

This is a word all students would be familiar with, and when asked to **explain** something in everyday conversation they would easily be able to oblige. However, in a test context you would be surprised how many students miss the mark.

For example, assume students were asked in a history assessment to '**explain** a famous historical battle' – there will be those who spent a lot of time researching, understood the content well, and wrote an in-depth account of where and when the battle occurred, the size of the armies, and the number of casualties. They may have included an impressive amount of detailed and historically accurate information, **BUT** if they did not provide information on **cause and effect** and address the **how and why**, they would have failed to meet the brief and would not be able to access the higher mark range. To properly answer the '**explain**' directive term, they would have had to provide additional information on why and how the battle started, as well as the outcome and impacts of the battle, thereby addressing cause and effect.

Let's take another example, this time relating to science. Imagine students were asked to conduct an experiment where two substances were mixed, causing a visible reaction, and then asked to '**explain**' this reaction. Students who limited their answers to visible observations (e.g. colour or state changes), even if they supplied accurate detail, would miss the mark. To score higher marks, students would have to **explain** the **how and why** or **cause and effect**. In other words, they would have to talk about which specific chemicals were reacting and the scientific reason why, including detail about the chemical process which took place to cause the observable reaction. They might also have to include a chemical or word equation to substantiate their answer.

In this science example, let's imagine that when the two chemicals were mixed, students observed that bubbles formed. An '**explain**' answer might look something like this:

"When magnesium (metal) and dilute sulfuric acid were mixed, bubbes formed because the metal reacted with the acid causing the production of H_2 gas which became visible as effervescence in the reaction vessel."

As a rule, regardless of the subject, if answers to **explain** questions lack detail around how/why or cause/effect then they are operating at an **outline** or **describe** level instead, and students who fail to answer the directive term correctly will be unable to achieve higher marks.

The key strategy for any **explain** question is to ask yourself the following:

- Have I mentioned the causes or reasons why something happened?
- Have I mentioned the end effect, and how those causes led to the outcome?

TIP

Adding the linking word 'because' ensures you are answering any explain question!

Looking back at our two examples you can see how simple and effective this is:

- 'The battle started **because**…'
- 'When the two substances were mixed, bubbles formed **because**…'

By inserting the word **'because'** into the sentence, it creates an automatic prompt to follow with an explanation. In fact, there is no way to avoid providing an explanation after **'because'**, since the very next words provide the reasons, addressing the how and why, or cause and effect.

Note: 'because' is not the only word that can be used to demonstrate cause/effect or how/why. It is an easy and effective one, but there are other alternatives that may be useful depending on what you are aiming to explain. Here are a few alternative linking words: due to, as a result of, consequently, since, therefore.

<u>Discuss:</u>

We use the word '**discuss**' in everyday language where it alludes to anything from a chat or conversation to a debate. In this context, however, the '**discuss**' directive term asks you to address a particular point, giving arguments for and/or against. Implicit in this is a requirement that students take a position and argue for this in their answer. If a student merely presents information without (a) addressing the point, (b) taking a position on the issue, and (c) presenting arguments for and/or against, they will not be able to access higher marks. In our experience, a sophisticated response usually includes both sides (for and against) as a 'discussion', which culminates in an overall point of view on the issue.

For example, if the question were:

"Discuss whether school uniforms should be abolished", the student should:

- Identify and state their position on the issue – this could be agreement, disagreement or an ambivalent view.
- If the student takes a view that uniforms **should** be abolished, they would provide most of their points and evidence in support of abolishing uniforms, while ideally also acknowledging some opposing points.
- If **disagreeing** that school uniforms should be abolished, the student would provide most of their points and evidence in support of keeping uniforms, while ideally also acknowledging the alternative perspective.
- If the student takes a mixed view, they would provide a balanced discussion with points for both abolishing and keeping uniforms.
- Whatever view is taken is fine as long as it's in the form of a discussion which weighs points for and/or against.
- The paragraph structure depends on the position taken – for instance a position of agreement would generally devote approximately two-thirds of the paragraphs to points **for**, and a position of disagreement would devote approximately two-thirds of the paragraphs to points **against**, while an ambivalent view would be balanced with approximately half devoted to each.
- The most important takeaway here is that marks will be won or lost on a student's understanding of what they must include in a **discuss** response – which is to **address a particular point/issue and give arguments for and/or against**, requiring students to pick and substantiate a point of view.

The uniform example was simple, so my recommendations above might be intuitive, but at school the commands are rarely one-dimensional, making it more challenging to address. An example of a more complex **discuss** question, such as one that may arise in an English study of Shakespeare, might look something like this:

"In William Shakespeare's play, Macbeth's downfall is primarily attributable to his ambition. Discuss this statement with reference to the play."

At this stage, we won't focus on unpacking the complexities of this particular question; I rather want to illustrate that students answering this prompt would need to pay close attention to the directive term '**discuss**' when planning their response. A higher level **discuss** response requires students to address Macbeth's ambition and clearly state their view regarding the role this played in his downfall. A sophisticated response might take a blended approach, positing that while ambition was indeed a significant contributor to Macbeth's downfall (argument 'for'), there were also other factors which catalysed his downfall (argument 'against'). This student would provide, analyse, and weigh evidence both 'for' and 'against', leading to a well-reasoned conclusion. There are numerous ways to answer the question, and this is only one approach, but the key to success is ensuring the answer meets the demands of the directive term **'discuss'**.

Analyse:

Analyse is a term you will encounter often across many subjects and can be tricky to address, being somewhat vague. The subtext for **analysing** is to '**understand and communicate how individual parts relate to each other to create meaning**'.

'Analyse' is used frequently in English throughout high school, as students are required from Year 7 to analyse various forms of literature such as poetry, prose fiction, plays, and films. There are specific techniques for analysing different question types and subjects, which I will shed light on later, but for now let's consider an English example. Imagine a student is asked to "write a paragraph **analysing** a scene from a film". If the student focuses their writing solely on describing what is happening in the scene, the characters, the plot, and pointing out film techniques such as music, lighting, and camera angles, they will likely get a disappointing mark. No matter how long and detailed the paragraph is, or how sophisticated the vocabulary is, they will be unable to access higher mark ranges because they have provided the incorrect level of response. Such an answer is aligned with a **'describe'**

level directive rather than an **'analyse'** level directive, which demands that the student communicate **HOW those individual aspects INTERACT to CREATE MEANING.**

In other words, although it's necessary to mention 'dim lighting' and 'sombre music', this will only attract marks if the student analyses **how this creates meaning**. So, the student would need to say something like: "The dim lighting and sombre music create a melancholic atmosphere to reflect the sense of sorrow experienced by the grieving protagonist." It's therefore the insight and attribution of **meaning** that truly answers the terms of the question.

Analyse is also frequently applied in science as students may be asked to conduct experiments, record results, or interpret second-hand data. When asked to **analyse** data or results, it would be insufficient to merely reference a table of measurements or observations. To access marks, students would need to conduct analysis communicating **the meaning of the results**. They would thus need to analyse the results in relation to the question, considering the aim, hypothesis, variables, and other experimental elements, showing an understanding of what the results 'mean' and the 'implications'.

CAUTIONARY TIP

Sometimes 'analyse' questions don't contain the word 'analyse', so watch for questions including the word 'how'. A 'how' question is often an analysis question in disguise.

For example, if the question is:

"How does the director of the film utilise visual and auditory effects to convey meaning?"

The student needs to know that they are being asked to **analyse** and must apply the same rules outlined above to effectively answer the question.

TIP

Use 'linking words' in your sentences to prompt analysis.

Similarly to an **'explain'** response, and depending on the question, you could use linking words such as **'because'** or **'as'** to prompt your analysis.

Also, using more specific linking words like **reveals, conveys, reflects, symbolises, demonstrates, shows, illustrates, highlights, suggests, evokes**, etc. prompt you to address the aspect of meaning. By completing the sentence, you are forced into communicating meaning. Let's revisit our example of the film analysis:

"The dim lighting and sombre music create a melancholic atmosphere **to reflect the sense of sorrow experienced by the grieving protagonist**."

If a student ended the sentence after the word '**atmosphere**', the analysis would be correct but incomplete. Adding the words **'to reflect'** forces students to explicitly address the meaning that has been created. It's this meaning that is necessary to fully answer an **analysis** question.

<u>Compare and contrast:</u>

These can be tricky terms, which catch lots of students out, so pay close attention!

To **compare** is to **show similarities AND differences between things**. Students sometimes lose marks unnecessarily by forgetting that high-level responses require **both** aspects to be addressed. This type of question can apply to a range of subjects. In English, for example, students may be required to **compare** two texts, such as two poems, a poem and a film, or a novel and a play. **Compare** questions could similarly arise in science, with students asked in biology to

compare animal and plant cells, or in chemistry to **compare** the results of two experiments. In all cases, students are strongly advised to include **both similarities AND differences** (with at least one of each depending on the mark allocation).

Students often confuse **compare** and **contrast**, but they require different responses.

To **contrast** requires students to focus solely on **differences**, and similarities would not need to be addressed.

NOTE: when answering **compare** or **contrast** questions you are required to look at two or more things in relation to each other. It's therefore crucial to **link** your points or ideas to each other, rather than dealing with them in isolation. This allows the marker to easily see that you are addressing the directive term.

Let's take another look at our biology example above to illustrate the point:

Imagine the question asks you to **compare** animal and plant cells. If you simply state features of animal cells in one sentence and features of plant cells in another, you will probably not access full marks, even if the information is accurate. For **'compare'** questions the marker needs to see that you are talking about one thing **in relation** to another. So, it would be advisable to integrate the information, using appropriate comparative language and linking words to clearly show you are addressing the directive term. So, you might say something like:

"Animal and plant cells are **similar**, **both** having a nucleus, cell membrane, cytoplasm and mitochondria. **However**, they **differ** because plant cells have a cell wall outside the membrane, giving them a rigid structure, **while** animal cells lack a cell wall making them more flexible."

You can see here that we addressed similarities and differences, integrating the information to highlight the comparison. If the student had simply stated that animal cells have a nucleus, cell membrane, cytoplasm, and mitochondria, and then separately stated that plant cells have a nucleus, cell membrane, cytoplasm, mitochondria, and a cell wall, they may not have achieved all the marks. While the information is scientifically correct, it has not been answered in a comparative way.

To assist you in addressing the comparative term and signposting this clearly for the marker, you should try to include the following **linking words** where appropriate:

Words that demonstrate **similarities**: both, similarly, likewise, also

Words that demonstrate **differences**: however, whereas, conversely, although, while, differ/differing, in contrast.

While relativity is equally important for **contrast** questions, remember you will only address differences, using linking words that demonstrate those differences.

Note: If a question asks you to **'compare and contrast'**, don't worry – this simply commands the inclusion of similarities and differences as in a **compare** question.

TIP

Using linking words helps you AND the marker.

Linking words act as prompts, ensuring you address the directive term, but equally important is their role as signposts for the marker, enabling them to easily and quickly identify that you have answered the directive term so they can award marks.

Assess and evaluate:

These terms are closely related, although subtly different. They both involve higher level directives, requiring a nuanced approach to addressing the question properly.

For both terms, students must **make an explicit judgement** to accurately meet the brief.

As outlined previously:

- **Assess** questions require you to judge the value, quality, results, or extent of something.
- **Evaluate** questions require a similar judgement although the judgement is made against specific defined criteria.

Evaluate and **assess** are often used in history and English questions, although they also appear in other subjects. Their subtle distinctions are better explained with actual examples, so I will show you how these types of questions may present, demonstrating exactly **how** to answer them to score top marks.

In history for example, the question may ask:

"**Assess** the value of source A to historians studying the technology used in World War I."

- This '**assess**' question is asking students to make a judgement about the overall value of the source.
- In this instance, a student who simply outlined or described the technology used in WWI would not achieve top marks, because the question requires an assessment of value, not simply an outline or description.
- To access higher marks, the student would need to explicitly state **how valuable** the source is and then follow up by **substantiating** their view.

- An example response might start as follows:
  "Source A is **highly valuable** to historians studying the technology used in World War I, **because** it successfully illustrates the development of communication technologies, instrumental to trench warfare…."

- You will notice a few things from this response:
  - o A clear assessment is explicitly made in the first line, establishing immediately that the source is highly valuable.
  - o The student includes the linking word **'because'**, prompting the evidence and explanation which substantiate their assessment of the value of the source.
  - o The assessment is followed by evidence of specific technology and its significance in WWI.

In English, for example, the question might be:

"**Evaluate** the effectiveness of imagery in creating a sense of happiness in the reader."

- This '**evaluate**' question asks students to make a judgement about the effectiveness of imagery in achieving a specific purpose. In other words, it is not asking how effective the imagery is overall, but rather how effective it is in relation to the defined criterion of creating a sense of happiness.
- In this instance, a student who focused on simply identifying the imagery would not achieve top marks, and an analysis of the imagery in a general sense would also not score well, because the question requires an evaluation of the effectiveness of the imagery in achieving its specific purpose.
- To access higher marks, the student would need to explicitly state whether the imagery was effective (or not) in creating a sense of happiness AND they would need to provide examples of imagery as well as reasons for how the imagery created (or failed to create) a sense of happiness, depending on the evaluation.

- An example response might start off as follows:
  "The composer's use of natural imagery is **extremely effective** in creating a sense of happiness in the reader, **as** he describes "bright florals" and "brilliant sunshine", emblematic of joy and upliftment...."

- You will notice a few things from this response:
  - A clear evaluation in relation to the specific criterion is explicitly made in the first line, stating that the use of imagery is **'extremely effective in creating a sense of happiness'**.
  - The student includes the linking word '**as**', prompting the evidence and analysis needed to support the evaluation.
  - The evaluation is followed by evidence of specific imagery, and analysis of how this conveys meaning in relation to the question.

An **evaluate** question could also be asked in a different and more subtle way, for example, an English question might ask you to:

"**Evaluate** how the composer of your text has represented emotions."

- In response, you would need to include specific evaluative language to address the directive term intentionally, and your answer might be something along the lines of:
  "The composer of the text has **skilfully** represented emotions through the use of ..."
- When asked to evaluate, you should always try to include words such as **skilfully, effectively, masterfully**, etc. to ensure that your evaluation is clear to the marker.

Bonus TIP

When a broad term is included in a question, unpack it into something more specific.

For example, in the previous point the student is asked about the **'representation of emotions'**. This is broad, and in most cases the text wouldn't be dealing with **all** emotions, so when answering, students should identify the specific emotions addressed within the text (e.g. sorrow, anger, joy) and evaluate on this basis.

CAUTIONARY TIP

Sometimes assess and evaluate questions do not contain those specific directive terms!

Instead, a question may ask: "To what extent...?"

This is really an **assess** or **evaluate** question in disguise, so it's important to know that if you encounter a question that asks: '**To what extent**...?' you should address it in the same way as **assess** and **evaluate** questions. Here are some examples related to English and history:

1. Shakespeare's *Romeo and Juliet* is primarily a play about love. **To what extent** do you agree with this statement?
2. **To what extent** did societal grievances contribute to the fall of the Romanov Dynasty?

Note:

- Both these questions require the student to make a clear judgement.
- The judgement will usually conform to one of three options:
  - A great extent
  - A moderate extent
  - A limited extent.
- The **'extent'** judgement should be **EXPLICITLY stated upfront** for the marker to easily recognise – in fact it should be in the first line!
- Whether a student chooses 'great', 'moderate' or 'limited' is less important...

- …what's more important is providing evidence and analysis to substantiate the chosen judgement and sustaining a coherent argument throughout the entire answer.
- If the judgement is 'to a great extent', the evidence and analysis will be heavily weighted towards substantiating agreement with the question/statement.
- If the judgement is 'to a moderate extent', the evidence and analysis will be balanced with points of agreement and points of disagreement.
- If the judgement is 'to a limited extent', the evidence and analysis will be heavily weighted away from the question/statement, supporting the opposing point of view.

Note: This type of oblique question arises often in HSC exams.

TIP

Include linking words like 'because' or 'as' wherever you can.

I have already stated this a few times for different directive terms. My repetition is intentional because this simple technique will assist in answering a variety of question types. The linking word prompts you to explicitly address the evidence and explanation, which are both necessary to substantiate your point. The inclusion of these linking words can be a grade maker.

Justify:

The **justify** term is quite self-explanatory and requires students to substantiate an answer or conclusion. In other words, they must provide reasons for the answer they have given. We have often encountered this in science exams, where students may be asked a simple question with an adjunct to **justify** their answer.

Some simple possible examples follow:

- At what time in the graph was the car stationary? **Justify** your answer.
- What type of rock is shown in the diagram? **Justify** your answer.

There are two common **pitfalls** with this type of question:

- First – students must know **how** to justify (or give reasons for) their answer, and this will often involve an explanation of sorts.
    - Once again, using the word **'because'** in their response will help guide them towards their reasons.
- Second – sometimes students are so focused on getting the correct answer that they forget to respond to the second part of the question, which asks for the justification. This is an easy way to squander marks but can be avoided by making sure to read and answer **ALL** parts of the question.

<u>Synthesise:</u>

'Synthesise' is a higher order directive term, usually encountered in the senior years as it requires a more sophisticated level of response. The directive is to select and combine information from multiple sources, **integrating** this to create a cohesive response. The tricky part is that this term can be combined either explicitly or implicitly with other question types, using directive terms such as analyse, compare, assess, and others.

Any time you are asked to **compare different texts,** or **refer to multiple sources**, this should be a flag that you are required to **synthesise** information from multiple places, weaving it together to make your point, whether this is in the context of assessing, comparing, analysing, etc.

An example of a synthesis question in English might be:

"Compare how the composers of Text 1 and Text 2 convey insight into the complexities of emotion."

- This is quite complex because it requires students to address multiple aspects:
    - They must **compare** the two texts, addressing both similarities and differences
    - They must **synthesise** the analysis, weaving refences to both texts throughout
    - All of this must be done while **analysing** the overarching question of 'how the composers convey insight into the complexities of emotion'.

An example of a synthesis question in history might be:

"Explain the changes in technology used in World War I. Integrate relevant evidence from Source A and Source B and your own knowledge."

- This is also a layered question, requiring students to address multiple aspects:
    - They must find relevant evidence from both sources plus their wider knowledge.
    - They must **synthesise** the information, weaving in refences to both sources and external information throughout.
    - This must all be done while **explaining** 'the changes in technology used in WWI'.

TIP

- **If you see the word 'integrate', this indicates a need to synthesise.**
- **If you are asked to reference multiple texts or sources, then synthesis is required.**
- **While synthesising, don't forget to also apply the relevant rules for any embedded directive terms e.g. explain, analyse, compare, assess, etc.**

***

All this information around **directive terms** may seem like a lot to digest, but it isn't very complicated. It's intuitively quite easy to understand and apply, and with **practice** it becomes second nature. You will note I say, 'with practice', because as we have established, skills are built over time with practice. I would consider these directive term skills to be some of the most crucial to diligently practise, because an ability to successfully master their application is, in my opinion, one of the most significant differentiators between those who 'hit the mark' with their answers and those who don't. This skill is required (and tested) in every subject from the start of Year 7 until the end of the HSC, so mastery will create academic strength and advantage in every subject. Conversely, weakness will adversely impact you throughout.

By way of example:

In a recent English Advanced HSC language paper, one of the questions asked students to, '**assess** how the author conveyed a certain aspect of the human experience'. (I have paraphrased the question because the exact wording is immaterial). The important part is that the marking guidelines stated that students who answered the question **without addressing the directive term**, failing to include an **explicit assessment** or **clear judgement**, would not be able to access marks above 60%. However, students who added even a single evaluative word to their response conveying a judgement, such as **'effectively'** or **'skilfully',** would have access to the top mark ranges.

So, to be clear:

- If a student had said: 'The author conveyed this aspect of the human experience by…' they would have been able to get a maximum mark of 3/5, regardless of how excellent their analysis was.
- However, if a student had said: 'The author **SKILFULLY** conveyed this aspect of the human experience by…' they would have been able to potentially score 4/5 or 5/5 (depending on the overall quality of their answer).

In this example, the inclusion of a **single evaluative word** could have taken the response **from a 60% to a 100% mark**, while overlooking the directive term would have cost valuable marks, no matter how accurate or sophisticated the rest of the answer was. This example is a simple illustration of the importance of addressing the directive term, and the higher the year group, and the greater the mark allocation, the bigger the consequence for failing to address the directive term. (Note that students would also have had to identify and address the specific human experience mentioned in the question as part of their **assess** answer.)

The **good news** is that once you get the hang of working with directive terms, you will automatically respond to questions in the appropriate way without having to spend valuable time deciphering what is being asked and what type of response is required. Establishing this skill early frees up cognitive load later, allowing you to spend more time adding nuance and sophistication to your responses.

TIP

Don't assume there will only be one directive term in a question – there may be more!

It's great if you can immediately spot a directive term…but keep looking in case there are others. Often, there is more than one directive term in a question, and if you forge ahead with the first, you may miss other vital parts of the question, thereby forfeiting marks that could have easily been achieved.

TIP

Check all parts of the question for directive terms!

Sometimes a question may have two or more parts, for example, it may have a stimulus, prompt or quote, with a question beneath it. Pay attention to all parts so you don't accidentally overlook a directive term.

> **TIP**
>
> **Highlight the directive term(s) in a coloured marker or underline them in pencil as a starting point in any question you encounter – whether this be an assessment notification for a large project or an individual exam question.**
>
> **And – If you don't immediately see a directive term, check if it's there in disguise.**

Highlighting the directive term(s), thereby making a conscious note of **what is being asked**, is a vital step towards ensuring you know what is required **before** starting work on it. This is essential for academic success, because if you know where you are headed from the start of any question or task, you are more likely to stay on track, answering the question correctly. In larger tasks, it will help keep you focused on the parts of work which are most valuable and relevant, preventing you from straying into territory that uses precious time without adding value. It helps to keep you working smart, ensuring that you are optimising your work time, staying within the most valuable 20% of effort, and giving yourself the best chance of winning marks in the most efficient way.

***

We have established that answering a question correctly involves first identifying what is being asked and then accurately responding to the directive term, but questions are rarely short, straightforward, and one-dimensional with a single directive term. More often, especially as you progress into senior years, the questions become long, convoluted, multi-part, and can contain more than one directive term, sometimes in disguise. It's easy to get lost in the question and consequently it can be a struggle to figure out what parts to zone in on when responding. Identifying discrete components of the overall question (beyond the directive terms) and being confident in answering **all** of them is a necessary skill which is acquired over time with practice. So, with this in mind, let's look at some strategies to take the process to the next level.

Techniques for answering BEYOND directive terms – and the all-important 'ATQ'

Before we delve into techniques for unpacking complex, multi-layered questions, let me introduce you to 'ATQ', which stands for **Answer The Question**.

We used to see this curious marking term scattered in red pen throughout various pieces of writing in the early years, and it took a fair while to work out that this was an abbreviation commonly used by teachers to flag parts of the work that were not deemed to be adequately 'answering the question'. This was highly valuable feedback once we understood what the acronym meant ☺.

On the surface, **answering the question** seems like it should be easy to do, and students (and parents) often fall into the trap of assuming thorough subject knowledge is all that's needed. This is, however, a dangerous and misguided assumption. It's entirely possible for a hardworking (and smart working) student to accumulate extensive knowledge and thorough understanding of the content and still get disappointing marks.

TIP

Thorough knowledge of the subject content is not enough!

Having subject knowledge is a necessary condition for academic success, but it's not enough, and only translates to high marks if the knowledge is **applied** in the right way for the specific demands of the question. In other words, students will almost never be asked to write down all the information they can remember on a given topic. Most questions they encounter will ask for an **application** or **interpretation** of the knowledge. This means rote learning and regurgitation of generic content will likely lead to disappointing marks. The students who earn higher marks will be those who are adept at figuring out what the question is asking at a **conceptual** level, and then skilfully **selecting** and **adapting** relevant parts of the information accordingly.

The skill of unpacking and answering questions can be challenging to master, because questions vary enormously. However, there are teachable techniques that can be applied in almost any scenario, so it's worth taking time to properly get to grips with this, because it becomes a great differentiator when it comes to winning marks.

The first part of this goes back to ensuring you identify and address the directive term, giving the correct level of response … but answering the question goes beyond that.

When faced with questions which are convoluted, containing multiple components that need to be addressed within the same response, it can be difficult to know where to start. In fact, even once the response has been written, it can be hard to work out whether you have adequately answered the question. In our hard-learned experience, the humanities subjects, especially English, are notorious for creating multi-layered questions which can throw even the most capable students.

Before embarking on the strategies for addressing complex questions, I will take a few steps back to share something that has become a bit of a joke in our family. Right before every exam, no matter the year level or subject, I only ever said one thing to my children: **"Read your questions properly."** This eventually evolved into: "**Read your questions properly all the way to the end**." I'm sure that as you read this you'll be thinking 'that's so obvious'... and you are right, it should be obvious, but I cannot count the number of times marks were needlessly lost in tests and exams for the simple reason that, under stress and time pressure, they either did not read the question properly or did not read all the way to the end.

The bad news: this mistake is often made by students, including high achievers.

The good news: it's very easy to address and improve immediately once you are aware and make a conscious effort.

TIP

- **Focus, read carefully, and absorb the question properly.**
- **Read the whole question, including secondary parts.**
- **Do not make assumptions about how the question will end.**

This might be one of the most valuable tips I can give regarding exam technique, because under time pressure and stress, it's easy to fall into the traps of glossing over the question quickly and superficially without absorbing all relevant details, or reading part of the question and forgetting about secondary parts, or even assuming (potentially incorrectly) how the question ends. The danger is unnecessarily losing marks by misinterpreting the question or providing an inaccurate or incomplete answer, despite having both the skills and knowledge. This careless waste of marks resulting from a failure to properly ATQ can be very frustrating, but if you are aware of how to avoid it in future, it becomes an easy fix.

TIP

Underline, circle or highlight key parts of the question or accompanying instructions – referring to these during and after completing the question to ensure you stay on track.

Reading over a question is passive, but as soon as you physically engage with it by circling, underlining or highlighting, it becomes an active exercise, meaning that your brain is primed to better absorb the requirements.

- This subliminally guides you towards addressing key demands in your answer.
- It makes the crucial parts of the question (and any instructions) stand out, so you are more likely to focus and less likely to overlook or forget them.
- It makes it quicker to check your work before moving on.

For example: imagine a maths question which asks you to **calculate** a distance in **metres AND convert** your final answer to **kilometres**. You may know how to do the calculation and get the correct answer, but if you forget the final conversion, marks will be lost. If you highlight metres and kilometres, you may still forget to convert if you are rushing, but once you have finished, if you were to glance back at the highlighted words, you would be more likely to catch and remedy the mistake to rescue the marks.

Another example: imagine a science question which asks you **to what extent** an experiment is **reliable**, and to **justify** your answer. You may correctly answer that the experiment is reliable, but if you neglect the second part of the question and don't provide your reasons/justification, you will sacrifice marks. Highlighting the word 'justify' helps draw your attention to the second part of the question and serves as a reminder when you check over your work.

- You will notice that both those questions contained two parts with directive terms, including one in disguise.
- You will also notice that I bolded the key parts of the questions
- Referring back to the **bold** words…
  - o 'calculate, metres, convert, kilometres' – in the first question
  - o 'to what extent, reliable, justify' – in the second question

  … in both cases, the key words stand out making it clear what needs to be done, and making it quicker to check, because you don't have to re-read the superfluous parts to establish whether you have answered the question.

The reality is that it's only possible to answer questions properly (even simple ones) if you have fully read, understood, and engaged with them. Questions will contain all the necessary clues, values, and features for ATQ, so it's important to ensure that your solution or answer addresses all aspects of the required information.

This is a great starting point, but these questions are quite simple. So, what happens when you get a more complex or multi-part question?

TIP

Break down the question into digestible chunks.

When confronted with complicated or multi-part questions, the process is fortunately very similar! There is, however, one extra step which is to first break down the question into more manageable parts, and then for each part apply the same principles we have already talked about. Let's look at an example – imagine you are given a history assessment with the following question:

"Identify a significant event in history, explain the main causes, and analyse the impacts. Your chosen event should have taken place between the years 1900 and 1960 (include a bibliography)."

The first thing to note is this one question is made up of three interrelated questions, each with its own directive term. There are also a couple of important instructions.

Step 1 is to break it into its individual parts and apply the techniques we have established above. So, it will end up looking something like this:

Part 1: **Identify** a **significant event** in history

- In simple terms you must **choose** and **name** an event.
- This event must be an objectively **significant** event in history.

Part 2: **Explain** the main **causes**

- You must first say **what** the **main causes** were.
- Then you must address **how** and **why** these things **caused** the event to happen.
- Your answer will be a cause-and-effect answer.
- Inserting the word **'because'** for each cause will help ensure you are answering at the correct level (for example: a main cause of **<insert event>** was **<insert cause> because <insert reason>**).

Part 3: **Analyse** the **impacts**

- You must first say **what** the **impacts** of the event were (i.e. what happened as a result of the event).
- You must then analyse **how** and **why** the event **caused** these impacts or consequences and what it **meant**.
- This **analyse** question further asks you to draw out the implications of the event.
- Note: in this part of the question the event becomes the 'cause' and impacts become the 'effect'.
- Again, you can use linking words to guide the response and attribute meaning.

Step 2 is to ensure that you take note of the instructions:

- Make sure your event took place between 1900 and 1960.
- Make sure you prepare and include a bibliography.

Step 2 is crucial because even if you skilfully meet the brief in answering the questions, if you choose an event from the wrong time period or neglect the bibliography, you will automatically lose marks.

Note: I have used a history example because the process can easily be illustrated, but the same process could be applied to almost any multi-part question in any subject.

Caveat: There are times when unpacking a question is not quite as straightforward – for example, in English, sometimes the question contains more subtle, intertwined layers that are hard to untangle, and I will address some of these later in the dedicated English section. Having said that, if we are talking about the 80/20 principle, applying these techniques will help you to meet the requirements in the majority of cases.

TIP

Take a step back and try to understand the question at a conceptual level.

While breaking questions into chunks and looking for directive terms is crucial, there are times when this won't be enough, and for more complex questions you will need to view the question as a whole, considering it at a conceptual level. In other words, you need to find the **big ideas** at the core of the question, because these are the ideas you need to address in your argument. If you can't identify the idea or concept, you won't be able to fully answer the question. Some examples of big ideas in English might be love, belonging, survival, identity, connection, or power. The list is endless, but this gives you an idea of types of concepts you may encounter.

TIP with a caution

Beware of questions containing a stimulus or a source!

Sometimes questions have distinct parts, perhaps containing a stimulus or source in addition to the question. In this case, the question usually requires you to reference, interpret, or analyse the information contained in the stimulus.

This can occur in any subject, for instance:

- In English, the question may be preceded by a text extract, quote, or image, and depending on the unit you are studying, this could be from a set work you are familiar with, or it could be an unseen source.
- In history, the question could be preceded by an image or text from a historical source, which would usually be previously unseen.
- In geography, the stimulus may be a map or written extract.
- In science, the stimulus could be a graph, data, or description of an experiment.

These are some examples of what you may encounter, but regardless of the subject or stimulus, the format of such questions will usually be quite standard. The source/stimulus will be presented first, followed by the question you have to answer.

The question itself could be anything – for instance:

- It could ask you to **analyse** an aspect of a stimulus
- It could ask you to **assess** the value of a source
- It could ask the **extent** to which you agree with a stimulus
- It could ask you to use the stimulus as inspiration or a **prompt** for a creative piece

OR

- In trickier scenarios:
  - o The question may not directly ask you to reference the stimulus – but beware, if the stimulus is there, **you must use it**!
  - o The question could be possible to answer independently without using the stimulus – this is a trap, **beware**! You still need to use the stimulus.

We have already talked in depth about identifying directive terms and answering questions in general, so this bit shouldn't pose a problem; however, the challenge of a preceding stimulus is that the question expects you to explicitly reference the stimulus, and your mark will be determined by your ability to engage with it thoroughly and convincingly throughout your answer.

Caution:

Students tend to make two common errors with stimulus questions:

- **They address the stimulus without answering the question**

  OR
- **They answer the question without referencing the stimulus.**

We have seen situations where students provide beautiful, sophisticated responses to questions using their own extensive knowledge, while completely forgetting to reference or quote the source. Conversely, we have seen students do a fantastic job of analysing the stimulus, referencing it thoroughly, but forgetting to answer the question.

Both errors are equally grave and will cost you dearly, because properly answering the question requires you to address both stimulus AND question in an integrated response.

A student who provides a high-level response will synthesise detail from the stimulus or source with their own knowledge to fully answer the question.

General guidelines for effectively handling questions with a stimulus/source:

- If there is a stimulus or source, you must **always** reference it in your answer.
- Read the associated question **before** reading the source or stimulus:
  - This saves time and increases accuracy because if you already know what question you have to answer when reading the stimulus/source, you can filter information, helping you quickly work out what is most relevant.
  - If you read the source/stimulus without knowing what the question is, you will need to re-read it carefully a second time to find the information.
- In your response, ensure you focus on clearly **answering the question** with consideration for the directive terms and our ATQ guidelines above:
  - If you don't answer the question, the battle is already lost before you even try to incorporate the stimulus.
- Make frequent reference to the stimulus or source, incorporating specific detail and quotes as evidence to support your argument.
- Avoid throwing in words from the stimulus if they aren't contributing to your argument in a meaningful way – this won't win you marks.
- Avoid simply summarising or paraphrasing the stimulus because this doesn't add value – it's the interpretation of the stimulus that scores points.
- Weave stimulus references into your argument throughout.
- The guidelines on how often to mention the stimulus differ depending on the length of response and the subject, but as a rule:
  - In an essay or extended response, reference the stimulus in every paragraph at least once, but ideally a few times to create balance and demonstrate sustained engagement.
  - In a short answer question based on a source or stimulus, you should be guided by the mark allocation, providing

approximately one piece of evidence from the stimulus for each mark, using the 'n' rule, where 'n' = the number of marks assigned to the question. It may be acceptable to use 'n-1' pieces of evidence but this depends on the nature of the question and the depth of analysis or explanation provided for each piece of evidence. For example:

  - o In a history source question for four marks, which asks you to **assess** the value of the source, you will likely gain one mark for making a clear assessment of value, with the other three marks coming from your evidence of value or limitations derived from the source itself, meaning you should reference the source **at least** three times.
  - o In an English language paper, if a 3-mark question asks you to **analyse** how a cartoon stimulus conveys a specific idea, you may gain one mark from answering the broad question, but to be safe you would ideally reference and analyse at least 2–3 visual techniques from the cartoon, such as colour palette, composition, wording, etc. to support your answer.
  - o In a science question for three marks, which asks you to **explain** what is happening in a graph stimulus, you will need to give an explanation using at least three pieces of meaningful evidence from the graph, clearly conveying cause and effect. You should reference specifics from the graph such as numbers, axis labels, trends, etc. to support your explanation.

- For questions which are part of a broader unit of study, you will be required to also draw on your own knowledge. Although it's essential to reference the source substantially, you will have to supplement with evidence beyond the source. If you derive your evidence exclusively from the source or stimulus without demonstrating broader knowledge, you will be marked down. For example, this would apply in cases where you are being tested on a unit in history, an English set work, the application of theory in a science module etc.

- For standalone questions such as those in an English language paper, it's fine to rely solely on what you derive from the stimulus, because these questions are not related to a prescribed text.

Note: In the senior years, you may encounter questions asking you to refer to two stimuli or sources. This is more complex as you need to answer the question with reference to **both** sources, but the same guidelines apply.

For example:

In an English language paper, you may be asked to **compare** how the composers of 'Text 1' and 'Text 5' convey a certain idea. In this case you would answer the **compare** question by integrating evidence of similarities and differences between the two text stimuli in relation to the relevant idea. The process is the same as for one stimulus, but you would need to draw from both texts in a balanced way. Once again, you would be guided by the mark allocation to determine how many pieces of evidence to include from the texts, and you would need to answer to the directive term.

TIP

Know when to STOP!

Before we move on from ATQ, I want to draw your attention to the flip side of the coin. We've gone into detail about what to include, but it's equally important to know what **not** to include. If you are clear on what the question is asking, and you have answered it fully, then STOP. You don't need to go beyond this. Repetition or superfluous information dilutes the clarity of your answer, wasting valuable time without generating extra marks. Adding information 'just in case' can be detrimental in an exam, because spending longer than necessary on one question leaves you less time to answer remaining questions, and you risk running out of time. Additionally, providing too much information may annoy the marker or create the impression that a student isn't sure of the answer, so lack of discernment may cost marks. As much as knowing what to say is a skill, there is also skill in knowing when you've said enough.

***

This section on decoding and answering questions has probably felt like information overload, but I firmly believe that mastery of these techniques is not only in the most valuable 20% of effort – I would say that ATQ is likely the **most crucial** of all the skills in generating academic success. No matter what other brilliant strategies are in place, if students can't **ANSWER THE QUESTION**, they won't be able to access marks.

We have spent significant time on strategies for interpreting and answering the **terms** of the question, but this won't get you all the way if you don't deliver the information **in the way the marker is looking for.** In other words, knowing **what** to say is extremely important, but knowing **how** to say it, is equally vital.

4. WHAT is the MARKER looking for?

The marker ultimately decides your results, so it follows that giving markers what they want is grade determinant and therefore crucial for academic success. This can be intimidating for students because it goes beyond content knowledge and application, requiring them to anticipate marker expectations. While this may seem like a challenging exercise in mindreading, let me reassure you that anyone can learn how to do this once they have been taught the techniques.

Let's assume by now you can successfully identify what the question is asking, and you have the knowledge and skill to adapt the information to ATQ …

The next step is understanding some of the considerations and learning the techniques for giving markers exactly what they want – which mostly fall into 'Key 2'…

How long and detailed should the response be?

I've already alluded to the fact that sometimes it's clear how much information is required, for instance, definitions are usually short, sometimes only a sentence. However, with terms such as **describe, explain** and **analyse**, it can be more difficult to gauge whether the marker expects three lines, a paragraph, or an essay.

> **TIP**
>
> **In any question, the mark allocation is a key indicator of the amount and depth of information you need to provide.**

It's important to pay attention to the marks assigned to each question, because this provides guidance as to the **length** and **depth** of the expected response. For example:

- A '1-mark' question usually only requires a few words, one sentence, or 1–2 lines of simple working, depending on the subject.
- A '2-mark' question usually requires a short response of 1–2 sentences, or a few lines of working, depending on the subject.
- A '3–5-mark' question usually involves a paragraph with greater depth and detail, including some evidence or analysis (depending on the subject). For example, an '**explain**' question in science or a short answer **'analysis'** question in English.
- A '10–12-mark' question will usually be classified as a mini essay, or a scaled-down version of a full essay, as described in the following point.
- A '15–20-mark' question demands an extended response with 3–4 body paragraphs with a minimum of three points or pieces of evidence and analysis in each, plus introduction and

conclusion. This occurs most often in humanities essay questions which command students to **analyse** or **assess**.

- In Year 7, a '20-mark' essay would require around 750–800 words, but in senior years you should aim for approximately 1,000+ words. It's not advisable to do more than 1,200 words because this wouldn't be achievable within most exam time limits.
- The general rule for science is to provide one piece of information per mark. For example, if you were asked to **identify** characteristics of a plant cell and the question was worth two marks, you would be expected to provide two distinct characteristics. If you were asked to **compare** animal and plant cells and the question was worth four marks, you would provide two similarities **and** two differences.
- For maths calculations, the mark allocation is also indicative of the amount of working required. A 2-mark question requires a few lines of working, while a 4-mark question would usually require a substantial amount of working.

> **TIP**
>
> **In addition to the mark allocation, pay attention to the amount of writing space allocated to each question in an exam paper.**

I am aware that this probably sounds simplistic, but the amount of writing space given in a paper is a clear indication of the amount of writing examiners expect for a response.

- If there are ten lines available to answer a question, you shouldn't write two lines.
- Conversely, if there are only two lines of space allocated for a question, you would not be expected to write a paragraph.
- However ... without creating a contradiction, I advise that if there is 'extra writing space' available on an exam paper, you err on the side of using **some** of the extra space. In our recent experience, this can be a bit of a 'trap' for students, because if most students use

the extra space, it undermines the efforts of those who don't. So, particularly for analytical questions, it's worth writing that little bit extra, as long as it's meaningful and not just waffle.

TIP

Stick to the word limit or time limit if these are stipulated.

Often hand-in tasks such as essays, information reports, or research projects, have a word limit or range. Similarly with a presentation, speech, or multi-modal/video, there is usually a slide or time limit. It's advisable to stick within the specified limits, as work done beyond this may not be marked or, alternatively, if a piece of work is substantially longer or shorter than the assessment guideline, this could attract a mark penalty. Students who don't adhere to the specified limits are needlessly throwing away marks.

TIP

Related to the last tip – Aim for close to the word or time limit on the first draft.

There is often a temptation to write more than required in the initial draft, intending to capture everything you **might** need and selectively cut down and refine later. This sounds good in theory, but we've learned the hard way that it's counterproductive and a huge time waster. The process of cutting down invariably involves multiple iterations of incremental 'cutting', creating more work in re-reading and re-wording, almost always taking longer than anticipated. Sometimes the trimming process consumes more time than preparing the initial draft. We have experienced scenarios where it's taken hours to cut 1,000 words down to 700. This defies the principles of good time management and the 80/20 rule, because the trimming, while necessary, is unlikely to add proportionate value, and that time could have been better spent on other important tasks.

Although it requires self-discipline and practice, a better approach would be spending a little more time creating the initial draft with greater discernment, staying in the most valuable 20% of work, and eliminating time spent cutting out excess. This would save time, which could be allocated to the most valuable 20% of another task.

What does the marking criteria or rubric demand?

I mentioned previously that many assessment notifications are accompanied by 'marking criteria', also known as 'mark schemes' or 'marking rubrics'. These are often placed at the end of the notification or issued separately, which means they are at risk of being overlooked or forgotten.

These rubrics can range from simple, prescriptive checklists to convoluted and conceptual requirements which may be difficult to decipher. They often contain guidelines of a more qualitative nature, addressing aspects such as coherence, structure, and sophistication of argument and expression, which extend beyond **what** is said and includes an emphasis on **how** it is said.

TIP

Pay attention to the marking criteria!
- **Adherence to the marking criteria wins marks.**
- **Ignoring the marking criteria is a grave error.**

In our experience, the **marking rubric is vitally important** as it explicitly outlines the criteria against which the student's work will be evaluated and marked. It spells out for students exactly **what** they must do and **how** they must do it to achieve top marks. You may have the knowledge to answer the question or complete the assignment, but if it's not expressed in the way the marker requires, or you focus on the wrong aspects of the task, or it lacks conceptual depth, marks will not be awarded. Close attention to the marking criteria therefore forms part of the most valuable 20% of work.

The rubric is usually divided into result 'bands', for example, if an assessment is out of 20: a mark of 17–20 may equate to an 'A', a mark of 13–16 may be a 'B' and so on. (These bands may differ depending on the school, subject, mark allocation, or task difficulty.) Regardless of the division points, each band will have guidelines detailing what markers are looking for when awarding marks, and the criteria that need to be met.

There is also differentiation **within** each band – for example: 17/20 and 20/20 may both be graded as an 'A', indicating students with both those marks satisfied the criteria to earn an 'A', but the relative sophistication or skill in achieving the criteria is acknowledged and rewarded. This means the marking criteria provide:

- Objective benchmarks, where a student's work is evaluated and graded against a predetermined performance standard.
- A level of subjectivity and relativity within bands, as students are being compared with others in the cohort.

Marking rubrics can vary substantially in complexity depending on the subject and year. Early in high school, if you follow the directions of the marking criteria it's not difficult to get high marks or even 100% in assessments. This gets more difficult as complexity increases later, but it is still achievable.

Note: The following examples have been created as a composite representation of the types of rubrics we have encountered over the years.

Example 1:

A 'simple' rubric could constitute a basic checklist specifying what information should be included to earn marks. In this case, students would literally need to go through the list, ensuring that all points are appropriately included. For instance, a Year 7 PDHPE assessment may require students to "create an information report on cyberbullying". The rubric could look something like this:

Marking Criteria – total marks available: 15	Marks available
• Define cyberbullying (1 mark if defined accurately, 0 marks if inaccurate or not defined) • Give two examples of what cyberbullying might look like (2 marks if 2 relevant examples given, 1 mark if 1 relevant example given, 0 marks if no relevant examples given) • Give two examples of online environments where it might occur (2 marks if 2 relevant examples given, 1 mark if 1 relevant example given, 0 marks if no relevant examples given)	5
• Give two strategies to deal with cyberbullying if you are personally experiencing it (4 marks for clear range of plausible strategies) (2-3 marks for one plausible strategy or two satisfactory strategies) (1 mark for poor or unclear strategy) (0 marks if no strategy is provided)	4
• Give two strategies to deal with cyberbullying if you are witnessing it (4 marks for clear range of plausible strategies to assist) (2-3 marks for one plausible strategy or two satisfactory strategies) (1 mark for poor or unclear strategy) (0 marks if no strategy is provided)	4
• Define the term 'bystander' (1 mark if defined accurately, 0 marks if inaccurate or not defined) • Define the term 'upstander' (1 mark if defined accurately, 0 marks if inaccurate or not defined)	2

Figure 7.1 An example rubric – information report

You will notice that this rubric is clear. It leaves little to the imagination, and students should easily be able to ascertain whether they have provided the correct level of information to earn marks. If they do what's explicitly asked, they should get full marks.

However, not all marking criteria are simple and prescriptive. Certain subjects are by nature less clear, and as you progress in high school, the rubrics become more intricate, multidimensional, and harder to navigate. Let's look at a more complex rubric.

Example 2:

A history marking rubric may be more complicated. It's unlikely to just be a straightforward checklist, instead often containing qualitative descriptors, which can be difficult to decode. Let's look at the following rubric which might apply to a history assessment task such as an extended response or essay.

Imagine that the assessment is an essay question asking students to "explain the causes of World War I". The marking criteria might look something like this:

Marking criteria	Mark/Grade
• Response is logical, coherent, sustained, well-structured and gives an explanation that answers the question fully • Response is supported by detailed, accurate and relevant historical evidence • Uses complex historical terms and concepts	A 21-25
• Response is sound and well-structured, with an explanation that attempts to answer the question • Response is supported by sound, accurate and relevant historical evidence • Uses appropriate historical terms and concepts	B 16-20
• Response is structured, with a description that only partially answers the question • Response is supported by adequate, accurate and relevant historical evidence • Uses some historical terms	C 11-15
• Response is simple, descriptive, generalised or incomplete with limited answer to the question • Response has limited accurate, relevant historical evidence • Uses limited historical terms	D 6-10
• Response is simple or incomplete with basic or no answer to the question • Response has basic or no accurate, relevant historical evidence • Uses basic or no historical terms	E 1-5

Figure 7.2 Sample rubric – history

> **TIP**
>
> **Ensure you understand every term in the marking criteria – even ones that may seem irrelevant or obvious.**

Before we unpack the levels of response that determine each grade, let's be clear on what each criterion in this history rubric means:

- Logical = your points make sense and the marker can understand each one
- Coherent = your overall argument is clear and the marker can understand
- Sustained = quality is consistent throughout the whole response (start to finish)
- Well structured:
    - o Formal essay structure with introduction, body paragraphs, conclusion
    - o Paragraphs have topic sentence, evidence, explanation, link to question
    - o Question is answered throughout (remember ATQ).
- Explanation = show cause/effect or how/why (remember the directive terms)
- Accurate = information is correct
- Relevant = information directly relates to answering the question
- Detailed = each point is fully explained and supported with evidence or examples
- Terms and concepts = formal historical language and ideas as taught
- Description = retelling of events with no reference to cause/effect or how/why and not really answering the question.

Once you understand the exact meaning of all the terms in the marking criteria it becomes clearer what differentiates an 'A' response from a 'B' response. It's important to note that many of these criteria go beyond simply identifying what the question is asking and answering it 'correctly', additionally demanding a certain level of detail, sophistication, flow of argument, structure of writing, and so on.

Let's break this rubric down – first you will notice that if you simply **described** the causes of WWI without addressing cause/effect or how/why, you would not access the A or B mark ranges and would automatically be limited to a maximum of grade C. This is because, as stated in the rubric, grades A and B require an **explain** level response. This refers to the demands of the directive terms, as addressed in the previous section, which are a critically important first step in decoding marking criteria. If you understand what level of response is required based on both the question and the marking rubric, it helps guide you towards fulfilling the criteria to achieve the top grades. However, this is not the only element to consider …

Let's assume you correctly identified the directive term and the appropriate level of response, and you **did** explain the causes … but you did not provide much detail and parts of your argument didn't make sustained sense. In this case you might get a B, but you probably still wouldn't access the A range, because there are multiple other aspects that need to be addressed or included. To achieve a mark within the A range, you needed to cater for **all** the criteria listed in the 'A' band descriptors. This means in addition to answering to the directive term, you would have had to write a detailed response with clear points, backed up by accurate and relevant historical evidence, using complex terms and concepts, culminating in a cohesive and sustained argument, all presented in a well-structured essay. Most of these criteria relate to the depth and quality of the response, beyond just being able to answer the question.

TIP

To determine what's needed for an 'A', examine the difference in words between the 'A' band criteria and the 'B' band criteria. The words in the A band that are absent in the B band, contain clues to what differentiates an 'A' from a 'B' response.

In this example, let's look at the specifics that differentiate between an 'A' and a 'B' (i.e. we are looking at the **differentiator** words which appear only in the 'A' band):

- Logical, coherent, sustained
- Answers fully
- Detailed evidence
- Complex terms/concepts

This means if you have met the baseline of all criteria in the 'B' band, and ADD the above criteria (or most of them), then you will jump from a 'B' to an 'A'. So, identifying what new criteria are introduced for each grade descriptor provides everything you need to know about how to jump from one band to the next.

***

TIP

Take time to go through the marking criteria thoroughly and more than once!

Review all available marking criteria **before** starting work on any assessment. If you know what the marker is looking for at the outset, it's easier to incorporate these elements from the beginning. It's more difficult and time consuming to backtrack, working them in retrospectively. It's advisable to check your work regularly against the criteria **throughout** the process, ensuring you stay on track, and again once the work is **complete**, to be certain you have ticked all the boxes.

CAUTION

Pay attention to what is not going to score marks!

We have established that it's vital to pay attention to what **is** stipulated in the marking criteria, but it's also important to note what **is not** stipulated. In a desire to perform well, it can be tempting for students to go above and beyond, spending time on aspects of the task which will not earn marks. For example, recalling our Year 7 cyberbullying task criteria, if a student spent time on elaborate infographics, it might look fabulous but would unfortunately result in time wasted because the marking rubric awards zero marks for graphics or creativity. This is a pitfall many high-achieving students fall into, even in the senior years.

Time and attention should be apportioned according to the mark allocation of the criteria. If creative criteria are secondary to answering the question properly, ensure you attend to ATQ first. In general, focus most on what will score you the highest marks.

TIP

Ask for feedback on your work whenever you can!

While working on an assessment you will find that some criteria are easy to address and self-evaluate – for example, it's quite straightforward to attend to structural and accuracy requirements. However, some criteria are less tangible and more challenging to self-evaluate. For instance, it may not be easy for you to judge whether your argument will seem logical and sustained to the marker. It's hard to be objective about your own work, and even though an argument or answer can feel clear in your own mind, it's important to test it on someone who can provide objective evaluation and feedback.

Seeking feedback with the aim of improving is part of developing a growth mindset. Most of the time teachers are willing to help students, providing feedback within the limits of what's allowed for an assessment. However, there are times when it's not feasible or possible to get teacher feedback, and in these cases parents, siblings, or peers can be equally valuable feedback sources, especially on the more qualitative criteria.

Note: getting feedback on your work relative to marking criteria will only apply in the case of hand-in tasks or practice tasks and won't usually be possible in class assessments beyond the preparation stage, as explicit marking guidelines may only be available during the assessment or after the task has been completed.

English

In decoding marking criteria, we have found English to be particularly challenging. The rubrics are less likely to have explicit checklists, and can be notoriously vague, convoluted and subjective. Given that English is compulsory, it's certain that every student will encounter this sooner or later, and it's arguably one of the most important areas to master, as English **WILL DEFINITELY** count towards your ATAR. If you conquer English mark rubrics, you will likely succeed in the subject, helping you overall. Unfortunately, the converse is also true. Now, let's look at an example ...

Example 3:

Referring to our earlier example of an English assessment notification, let's examine the rubric. As a reminder, the task is as follows:

"Interview a family member, friend or member of your community who has experienced a significant moment in their life. Use this as inspiration to create your own fictional short story about 'a significant moment in life'."

Marking criteria	Mark/Grade
• Demonstrates a **sophisticated** ability to compose an original short story with an authentic and engaging character, setting and complication with regard to a significant moment in life • Demonstrates a **sophisticated** portrayal of context and values of the character • Displays **sophisticated** control of language appropriate to audience, purpose and form	A 17-20
• Demonstrates a **substantial** ability to compose an original short story with an authentic and engaging character, setting and complication with regard to a significant moment in life • Demonstrates a **substantial** portrayal of context and values of the character • Displays **substantial** control of language appropriate to audience, purpose and form	B 13-16
• Demonstrates a **sound** ability to compose an original short story with an authentic and engaging character, setting and complication with regard to a significant moment in life • Demonstrates a **sound** portrayal of context and values of the character • Displays **sound** control of language appropriate to audience, purpose and form	C 9-12
• Demonstrates a **limited** ability to compose an original short story with an authentic and engaging character, setting and complication with regard to a significant moment in life • Demonstrates a **limited** portrayal of context and values of the character • Displays **limited** control of language appropriate to audience, purpose and form	D 5-8
• Demonstrates a **basic** ability to compose an original short story with an authentic and engaging character, setting and complication with regard to a significant moment in life • Demonstrates a **basic** portrayal of context and values of the character • Displays **basic** control of language appropriate to audience, purpose and form	E 1-4

Figure 7.3 Sample rubric – English

The first thing to note is that the criteria for each grade band are almost identical in terms of **what** must be included for the task – which is composing an original short story responding to the question, portraying the context and values of the character, and displaying control of language (which relates to the stylistic features, conventions and level of formality appropriate for the type of writing). There is a checklist of sorts, clarifying the objective elements which must be included, such as a character, setting, a story complication, a significant moment, context, and values.

However, if you pay close attention to the highlighted words, you will see the grade is determined, not just by the inclusion of certain aspects, but more importantly how **well** they are executed, and the sophistication of the response. For example, a student who does not include anything on the context or values of their character will get a poorer grade, while a student who includes insightful detail on both the context and values of their character with a high level of sophistication will get a stronger grade. While it's relatively easy to know if you have included elements of context and values in a binary sense, the more difficult part is figuring out whether you have addressed this to a **'substantial'** or **'sophisticated'** level. This is incredibly tricky. In full transparency we found that English marking criteria such as this are notoriously difficult to navigate, because descriptors such as **'substantial'** and **'sophisticated'** are not easily quantifiable, tending to be inherently subjective. While it's easier to distinguish at the extremes, the lines between grades, especially towards the top end, are blurry, and even teachers might have differing views among themselves regarding whether a response deserves 16/20 (high B) or 17/20 (low A). Over time you can develop somewhat of a feel for this, but it can be open to interpretation.

The good news is that marking criteria for English are reasonably predictable and consistent, so if you work this out in Year 7 you can build on the same skills and principles all the way to HSC. The complexity may increase, the questions will change, and the types of assessments may vary, but the principles remain the same. The skills learnt navigating English marking criteria are completely transferable, applying similarly for a Year 7 speech and a Year 12 unseen essay. Practice in this area has a large payoff, with benefits accumulating over time, translating to higher marks overall.

I am going to provide one more example of a Year 12 English marking rubric, representative of something you may encounter in an unseen essay exam question. It is generalised and could apply to any number of questions, which is why I have not alluded to a specific question in this instance. My intention is not to incite fear in younger students, but to demonstrate the transferability of skills, showing an end point, so parents and students can see what skills they are building towards.

Example 4:

Marking criteria	**Mark/Grade**
• Interacts skilfully with the question, demonstrating sophisticated understanding of human experiences in the prescribed text • Skilfully analyses how human experiences are represented in texts using detailed textual references from the prescribed text in response to the question • Presents a coherent, sustained, extended response with language appropriate to audience, context and purpose	A 17-20
• Interacts effectively with the question, demonstrating substantial understanding of human experiences in the prescribed text • Effectively analyses how human experiences are represented in texts using detailed textual references from the prescribed text in response to the question • Presents an organised extended response with language appropriate to audience, context and purpose	B 13-16
• Interacts adequately with the question, demonstrating some understanding of human experiences in the prescribed text • Presents a response using some textual references from the prescribed text • Presents a response with inconsistent control of language appropriate to audience and purpose	C 9-12
• Limited interaction with the question, demonstrating limited understanding of human experiences in the prescribed text • Describes aspects of the prescribed text • Presents a limited response	D 5-8
• Refers to the prescribed text in a basic way • Attempts to write a response	E 1-4

Figure 7.4 Sample rubric – English Year 12

You will notice some patterns consistent with our earlier rubrics:

- Many terms are familiar from previous rubric examples, demonstrating transferability of criteria across subjects and year groups.
- The differentiators are qualitative in nature (and thus more subjective).
- The 'A' band requires skilled and sophisticated expression and engagement with the question, along with analysis and detailed, sustained, coherent responses.
- The 'B' band has almost identical requirements to the 'A' band, but responses at this level would be **'effective'** and **'substantial'** rather than **'skilful'** and **'sophisticated'**.
- The lower grade responses are lacking in analysis, question engagement, insight, detail and consistency.
- It would be easier to move from a C to a B with some objective adjustments, but harder to move from a B to an A, and even harder to move from a mid A to a high A, due to the subtlety of elevation in criteria.
  Note: This links to the principle of diminishing returns discussed earlier – the closer you get to the top, the greater the required investment for smaller gains. Having said this, in our experience, English is worth the investment given its significant contribution to your end results and ATAR.

We have now spent quite some time specifically on English, but this advice is broadly applicable across subjects and years.

***

You should feel reassured in the knowledge that time spent learning how to decode and effectively address marking criteria from the start of Year 7 will build and compound over time to generate tremendous strategic advantage.

TIP

Actively seek out the marking criteria in every assessment notification and exam – sometimes it's not obvious, but if it's there, make certain you find it!

We have established that most hand-in or submission tasks will have marking criteria attached to the assessment notification, outlining what markers are looking for. There are also times when marking criteria may be given in advance of in-class exams, which is very helpful in guiding exam preparation. However sometimes marking criteria are not provided in advance, only appearing in the exam itself, for instance, in unseen English essay exams. Be aware that these rubrics might not look like the full examples we have shared here and instead may be represented as a couple of brief sentences declaring **"what you will be assessed on"**. This is certainly the case for some components of the English HSC final exams. Under time pressure, and with intent focus on the question itself, these marking criteria can be easily overlooked.

For example: In a recent HSC English Advanced final exam, the criteria outlining what students would be **assessed on** stipulated that they would be required to **'explain', 'analyse'** and **'assess'** in their response. The actual question students needed to answer was on the following page, and this question did **not** explicitly ask for an **'assess'** level response. Many students would have overlooked the criteria on the previous page and answered the question without making an explicit judgement in their response. This small, but significant omission would have compromised their marks.

It's therefore crucial to actively seek out marking criteria, making a deliberate mental (or physical) note to decipher and incorporate these criteria into your response. Focusing only on the **'what'** and ignoring the **'how'** can cost you valuable marks.

Remember that attention to the marking criteria is grade determinant, therefore falling into the most valuable 20% of effort.

<u>**TIP**</u>

Review the marking criteria after the exam too!

After assessments have been marked and returned to students, teachers usually provide model solutions, sample answers, and/or 'marking guidelines'. These contain specific information on what they were looking for when awarding marks. This is a highly valuable resource, and attention to retrospective marking criteria will often distinguish between good students and great students.

For many students, once the exam is done, marks are finalised, and they have absorbed any personalised feedback written on their exam paper, they file it away physically and mentally and move forward. This is a wasted opportunity and students who recognise the value of carefully reviewing past tasks against marking criteria gain an academic edge. Looking back is equally important as looking forward. This is a Quadrant 2 activity because although not urgent, it is very important for improving the quality of your future work. This is also a tangible way for students to engage with a growth mindset, focusing on learning and adapting to improve their outcomes.

Reviewing work retrospectively against marking guidelines provides the following insights:

- You understand exactly what the marker was looking for because they often break down and explain the **process** of mark allocation
    - o E.g. in a science question the marking guideline may show that an answer containing **x and y** was awarded two marks, but an answer containing both **x, y AND an interaction**

between the two, was awarded three marks – here you would learn the importance of including that interaction in future.
    - E.g. in a maths question, the marking guideline may show three specific bits of working that attracted marks, illuminating that even a correct final answer may only be awarded 2/3 if a specific line of working was not shown – here you would learn the importance of showing certain steps in your working rather than skipping steps or doing them mentally.
    - E.g. in a written response you may have said the right thing in the wrong way, or omitted a specific required term, and the marking guideline may show a model response – here you would learn **how** to express an answer in the way the marker requires and apply this to future questions.
- You gain insight into the mind of the marker to better understand how **qualitative** components were assessed – for example, in an essay exam the marking guidelines may provide general cohort feedback giving insight into what worked well in stronger responses, or what required improvement in weaker responses – this cohort level feedback provides concrete instruction on the specific subtle components that either earned or lost marks.
- Reviewing work against marking criteria, considering both individual and group feedback, allows you to **reflect** on what you can do better, putting **strategies** in place to refine your skills and improve your performance in future questions or tasks.
- You also gain **insight into what you did right!** Too often, feedback and self-reflection concentrate on what was wrong. While this is important for progress and improvement, it's equally important to know what you did right and what aspects of your work won marks. This vital part of the review process is often overlooked and may seem counterintuitive. You might think it's wasteful spending time on something you already know how to do correctly, but it's quite the opposite. Just because you did it right this time, doesn't mean you will necessarily do it

right next time. Taking time to absorb **why** you got it right helps you internalise this learning so you can replicate it in future.

TIP

For Year 12 – look at official HSC marking feedback and guidelines.

HSC markers provide marking guidelines, sample answers, and cohort level feedback every year after marking is complete! This often specifies what was present in stronger responses and what students need to improve on. This is extremely helpful to consider when preparing for your exams.

This little known, and incredibly valuable resource, is freely available to all students on the NESA website (www.educationstandards.nsw.edu.au). They provide 'HSC exam packs' with past papers and marking guidelines going back years. I highly recommend that all Stage 6 students look up their relevant subjects when preparing for HSC exams. Familiarity with what examiners look for is absolute gold and will not only assist with final HSC exams but recognising and working towards these guidelines will also help with internal school assessments.

***

The key focus of this section has been **how to answer the question** (ATQ) and **how to give the marker what they want**. You might have noticed throughout our examples of marking criteria that in addition to ATQ, an important part of giving the marker what they want involves the structure, expression, and conceptual depth of the response, especially for essays. So, let's unpack this a bit further.

How should the response be structured, and what elements must be included?

When structuring essays and paragraphs, there are certain conventions you need to adhere to. Markers look for these even if they are not explicitly mentioned in the marking guidelines. In meeting these requirements and creating a well-structured response, you will give markers what they are looking for, elevating your work and creating opportunities to score higher marks.

This is more relevant for subjects which require essays or extended responses, such as English, history, economics, and other humanities. As a rule, essay responses require an introduction, 3–4 body paragraphs (depending on the subject), and a conclusion. However, it's not quite so simple and each component has its own structural and content demands, so let's break this into digestible chunks:

<u>Part 1 – The introduction</u>

Your introduction should contain:

- An explicit answer to the question (ATQ)
- A clear thesis
- Context
- A brief outline of the main point you will make in **each** paragraph
- A link back to the question

We have already devoted a significant amount of time to the skills and strategies needed to **answer the question**, however, in an essay or extended response there is another vital component, which goes hand in hand with ATQ … the **THESIS STATEMENT**.

The thesis statement is essential for most essay subjects and usually consists of 1–2 sentences which sum up your **main argument** or **key idea**. This statement appears upfront in the introduction, and although it's instrumental in answering the question, it has an important function beyond ATQ. While the thesis is a crucial component of essays for all writing subjects, I will spend some time focusing on English, as this can be one of the most complicated thesis applications and, as we know, students will definitely be required to master this as a compulsory part of HSC.

Remember in the introduction of this book I mentioned there were some lessons we learnt the hard way, that we wish we had known earlier? Well, the importance of the thesis statement was one of those very hard-learned lessons. I will digress briefly at this stage to share an anecdote, because I think it's the only way to fully illuminate the vital importance of the thesis.

My daughter was in the first term of Year 8, studying Shakespeare's *Romeo and Juliet* in English. The assessment notification was issued, and the task was to be an in-class essay exam, answering an unseen question. It was the first English task of the year, and the weighting was 20% of the total year mark, which was significant. She had prepared thoroughly: annotated the play, created a comprehensive list of quotes, identified literary techniques, prepared analysis around numerous themes, and practised how she would adapt to answer various questions. She undertook the exam and felt ok about it (which is probably the best you can hope for in English, being such a tricky subject). The exam essay question had asked students something to the effect of 'how Shakespeare explored the impact of different types of love on characters throughout the play'. She had structured her essay well with the conventional introduction, three body paragraphs and conclusion. The essay was a good length, she had addressed three distinct types of love (familial, platonic, and romantic), devoting a paragraph to each, and had chosen appropriate characters to demonstrate the impact of each type of love. She had used the right number of quotes and conducted solid analysis for various characters. Overall, she felt she had done a pretty good job under exam conditions with an unseen question …

… Fast forward to a couple of weeks later when marks were returned. She was devastated to receive what she considered a very poor mark, not in keeping with her self-imposed high standards, and certainly lower than any mark she had ever received. What followed was significant distress and self-doubt. She read the marker feedback but still wasn't entirely sure what had gone wrong. Eventually she asked her close friend, also a strong English student, to read it and provide some insight. It took only a few minutes for her friend to diagnose the problem …

… there was no overarching thesis!

While each point and paragraph had analytical merit and answered the question, there was no binding thesis tying everything together. This was what we refer to as an 'aha' moment as we learnt (the hard way) the vital importance of a thesis, and the academic cost of an essay without one.

The thesis is the backbone of the essay. It's the glue binding all the individual components into a well-rounded and cohesive whole. Put another way, it's like having three separate ingredients, which alone are just ingredients, but when combined, create a tasty complete dish (hopefully). Although it sounds simple and logical, it can be challenging to create a good thesis, and even more tricky to create a good thesis that directly answers the question and can be sustained throughout an entire essay. This is something many students struggle with conceptually, even for hand-in tasks where they have time to think clearly. However, when adding time pressure and the stress of an unseen question in an exam, this can be extremely difficult to get right without having acquired and practised the proper skills.

Let's reconsider this *Romeo and Juliet* example, remembering that a **how** question is generally asking for an **analysis** level response. A simple overarching thesis could have been something like:

- Shakespeare explores the impact of different types of love **through the emotions and choices** of his characters, or

- Shakespeare explores the impact of different types of love to **cause joy or inflict suffering** through the experience of his characters, or
- Through the experiences of Shakespeare's characters, he explores the impact of different types of love to **be both beautiful and destructive**".

In these examples, the **thesis** is simple (appropriate for Year 8) and contains a broad **overarching conceptual insight** that **answers the question** and **binds** together all the arguments you will make in each of your paragraphs. All these thesis statements could lead into and support the exact three paragraphs my daughter had written. She would still have been able to address familial, platonic, and romantic love, using the same characters, quotes, and analysis. If she had used the first thesis, each paragraph would link to the broad idea of how love impacts emotions and choices, if she had used the second, each paragraph would link to the broad idea that love can induce joy and/or suffering, while the third would have each paragraph linking to the potential for love to be both beautiful and destructive. I have provided simple thesis examples to make the point, but with practice students would be able to create more nuanced thesis statements, and this is certainly something that would be desirable to build towards in the senior years.

<u>**TIP**</u>

Do not create a thesis so broad and generic that it conveys no meaning.

Many students fall into the trap of saying something generic like: "Shakespeare explores the impact of different types of love **through his use of literary techniques**."

While technically this is true – Shakespeare does absolutely use literary techniques to convey meaning – it is an empty thesis. It is bland and lacking substance, failing to address the **'how'** part of the question adequately.

Another common mistake made by students is dropping a thematic label into the thesis without taking this any further. In other words, simply stating that Shakespeare explores love or types of love would be meaningless and superficial. These types of thesis statements are unlikely to score great marks in an essay or paragraph response as they are limited and largely non-analytical. This leads directly to the next crucial tip …

> **TIP**
> **A great thesis must contain a conceptual insight.**

A strong thesis containing a conceptual insight will transform your writing to achieve top marks. This is the part of the thesis that deals with the meaning or **big ideas** conveyed by the composer. Instead of stating that Shakespeare explores the theme of love, it's necessary to talk about **what** exactly Shakespeare says about love, and **why** and **how** he says this. The conceptual insight is what takes your argument beyond a superficial retelling of the plot and broad generic statements, instead demonstrating a thoughtful and considered response. This is where you show understanding of the text, its themes and ideas, and its perspectives and lessons for the audience. This is where sophistication lies, and a clear conceptual thesis established in the introduction pays tremendous academic dividends, because it immediately highlights to the marker that you have a deep grasp of the textual meaning.

In our *Romeo and Juliet* example, the conceptual insights from our proposed theses would be that love can impact emotions and choices, or love has potential to cause joy and suffering, or love can have beautiful and destructive impacts. These insights reveal ideas relevant to humanity, going beyond the plot, conveying meaning and purpose.

An excellent thesis therefore requires at least three elements:

- It must **answer the question** (preferably using key words from the question).
- It must be an **umbrella** for your **key points**, binding them into a cohesive whole.
- It must have a clear **conceptual insight** incorporating the big ideas or themes.

***

From our perspective, we affectionately refer to the *'Romeo and Juliet'* incident as our own personal 'Shakespearean tragedy', wherein a VERY hard lesson was learnt. It is certainly one that has never been forgotten and has been consistently applied in all English endeavours since.

To close the loop on our thesis anecdote:

The following term, my daughter received the assessment notification for Task 2, testing their study of Brian Percival's film *The Book Thief* (2013). This time they were required to prepare a mini essay in the form of a multi-modal video presentation. I am paraphrasing, but the question was along the lines of: "In the film, *The Book Thief*, how does the director demonstrate the power of words?" In the past, before her hard-learned lesson, my daughter may have answered the question by simply saying: "In the film, *The Book Thief*, the director demonstrates the power of words through the use of film techniques." This is, in fact, how some students did answer the question. However, in grasping the importance of an insightful binding thesis, she created the following overarching thesis to ATQ:

"A significant theme in the film *The Book Thief* is that words have power to manipulate, control, and destroy free thinking. Percival prominently illustrates this insight in the book-burning scene by utilising film devices which support this meaning."

As you can see, she included all three elements by answering the question directly, creating an overarching thesis, and including a meaningful conceptual insight about the power of words. For this assessment, she was awarded 100%.

I must point out three things:

- The thesis alone did not win her top marks, and she would have needed to meet all the other criteria well, BUT **without** a well-formed thesis, she would not have been able to access the top mark regardless of how many other boxes she ticked.
- While this thesis hit the mark in Year 8, it would have needed further development and nuance if it was a Year 11 or 12 task.
- If we consider the three ideas within the thesis (manipulate, control, and destroy free thinking) as points A, B, and C – in this thesis they are analysed as three separate points (A+B+C), but in Year 12 a marker would expect an overall sophisticated argument to represent an interaction of the ideas (AxBxC) rather than a cumulative listing.

***

Having established the importance of a central thesis, here are a few practical <u>TIPS</u> for developing a strong thesis and ATQ for English

- **Highlight** key words/directive terms in the question that need to be answered.
- **Begin your thesis statement with a rephrase of the question** (or at least use key words from the question).
- **Use open-ended linking words like: 'because', 'by', 'to', 'as', etc.**
- **See what words flow from this and whether these can form the basis of your insight.**

This probably sounds quite vague and can best be illustrated by example.

If we use *The Book Thief* example again, imagine the question is:

"How does the director demonstrate the power of words?"

In developing a thesis which answers this question and contains an insight, it's best to:

1. **Highlight key words** from the question to unpack what you are answering for. In this case the word **'how'** is significant as it's a directive term in disguise, signalling that you need to create an **analytical** level response. The **'power of words'** is significant because this is the **conceptual** crux of the question. The word **'demonstrate'** is also significant because you have to give evidence of how the director **shows** his point to the audience.
2. Start by **rephrasing the question** as a statement. The thesis statement might begin with "the director demonstrates the power of words…". This automatically guides you to directly ATQ – in fact you can't avoid moving in that direction.
3. The next step is to develop the specific idea or insight about the power of words, and this can best be achieved by adding an open-ended **connector** (linking word) into your sentence, e.g: "The director demonstrates the power of words **to**…" By adding the word **'to',** you essentially prompt yourself to answer the question: 'the power of words **to do what**?'
4. Once you have posed this question, you are practically forced into an **insight**. In this case, from my daughter's interpretation of the film, the ideas that came to mind were **manipulate, control,** and **destroy free thinking**. Other students had different words and insights. There is no right or wrong, but a conceptual insight that goes beyond plot or technique to address the **big ideas** will always be a fantastic start to a strong conceptual thesis.
5. Once you have made it this far, you can then add a second **connector** such as **'by'**, which prompts you to address the **'how'** part of the question.

Putting it all together: "the director demonstrates the power of words **to manipulate**, **control,** and **destroy free thinking, by**… ". The last part will contain your evidence for exactly what scene and/or techniques he employs to demonstrate the power of words.

This was quite a straightforward example, but the principle is broadly applicable. If you reframe the question as a statement and try to extend or qualify the main idea(s) using appropriate connector/linking words as prompts, this will help to develop insights.

But what about more complicated questions which arise in senior years? This can be difficult, so let's look at a more challenging scenario. You might recall we talked at length in an earlier section about directive terms, unpacking questions, and recognising that there may be multiple parts or layers to address. We have now added the complexity of not only answering questions but developing a thesis with insights.

Consider the following question designed for a Year 11 assessment in the study of Shakespeare's play, *Twelfth Night*. (Note: familiarity with the play is **not** required and this is simply being used to illustrate a process):

'"Composers create characters who evoke empathy in the audience."

In response to this statement, analyse how Shakespeare uses form to shape meaning.'

As you can see there are multiple layers to this question. The first thing to do is apply the skills learnt earlier, breaking it into distinct elements that need to be addressed. So, let's first **highlight** the parts of the question that need to be answered:

'"Composers create **characters** who **evoke empathy** in the audience."

In response to this statement, **analyse how** Shakespeare uses **form** to shape **meaning**.'

Many students might find this initially overwhelming because there are five different elements that must be answered, and these need to be combined into one insightful thesis. To add even more complexity, there is a statement and a separate question, **both** of which need to be addressed. As you can see, it's helpful to break it down and highlight, because this guides clarity on what aspects to include.

The next steps are **rephrasing** the question, integrating elements of both the statement AND question, and adding '**connectors**' to prompt **insights**, for example:

'William Shakespeare uses the [form] of *Twelfth Night* **to** create [characters] who evoke empathy **as** … [ask yourself why they evoke empathy]. **As** the play [ask yourself what unfolds in the play], Shakespeare [shapes meaning – ask yourself what meaning].'

As you can see, this creates the structural foundation for an effective thesis, but with the added complexity of this question you will notice one more thing that needs to be done:

General terms must be transformed into meaningful specific terms. In other words, the question mentions 'form', but for the thesis to have substance you would not just use the generic word 'form'. Instead, you would substitute in the **specific form** used in this play, which is 'tragi-comedy form'. Similarly, you would not use the general term 'meaning', and you would instead elucidate the actual meaning that Shakespeare is intentionally shaping. You might even want to be more specific with 'characters' and instead talk about 'complex characters'. It is this process that creates the conceptual insights, addresses the big idea, and adds substance, facilitating effective analysis.

Let's now add in these qualifiers and put it all together to create an insightful thesis:

William Shakespeare uses the **tragi-comedy form** of *Twelfth Night* to create **complex characters** who **evoke empathy** as they **suffer the**

consequences of their social transgressions. As the play **shifts from its comedic exposition to its tragic end**, Shakespeare **warns of the harmful ramifications of defying social boundaries.**

I have **bolded** some of the key words from the question so you can see how they are integrated in the rephrasing. Note that not all key words are there because if you are answering the question and addressing the elements of 'analysis' and 'meaning', you won't need to insert these actual words.

I have underlined the connective words which prompt insights, and I have highlighted the **qualifiers** and **insights.**

Even though this is a more complex question with multiple parts, you can see how application of the same formula can work extremely effectively to guide you to developing an insightful thesis.

To recap – the general formula for creating an insightful thesis has the following steps:

- Highlight key words from the question and prompt
- Begin your thesis statement as a rephrase of the question (or at least incorporate key words from the question)
- Use open-ended connector words like: 'because', 'by', 'to', 'as', etc.
    - words that flow from these become the basis of the conceptual ideas/insights
- Replace generic terms with the specifics, and add qualifiers wherever possible
- Replace words like 'meaning' with an actual interpretation or key message which aligns with the big ideas or concepts
- Ask yourself questions about 'what' or 'why' along the way to clarify your insights and provide substance.

Caveat: The steps outlined here will work for many scenarios, but there will be times when certain questions don't lend themselves to this exact formula. If you are well practised and have a good grasp of the characteristics of a strong thesis, you will be able to adapt accordingly. The more confident you become with the techniques of thesis construction, the more you will be able to stray from the formula and still create a sophisticated thesis to achieve top marks.

Note: I have intentionally used examples from Shakespearean set works because most schools teach and examine Shakespeare across the year groups, and this is an area of difficulty for many students. I believe if students can master essay techniques for Shakespearean texts, they should be able to apply this to any other unit of English.

Although I have focused specifically on English in these examples, the general rules of thesis construction can also be applied to other essay subjects.

CAUTIONARY TIPS for the THESIS

- Do not make thesis statements overly complex or convoluted, because it's easy to get tangled up in them; they should be clear and easy for the marker to follow. Very complicated thesis statements can be confusing for you and the marker, which can compromise you on other aspects of the marking criteria, costing you marks.
- Make sure you can substantiate everything you claim in the thesis. You might create a very impressive and sophisticated thesis, but if you can't back it up with evidence in a sustained way throughout the essay, it will cost you marks.
- Ensure that the thesis umbrella is broad enough to cover the main points of each paragraph that follow. If your later arguments go beyond or don't align with your overarching thesis, this disconnect will also cost you marks.

- Your central thesis does NOT need to be revolutionary or unique – trying to be niche can make things harder than they need to be and can count against you if the teacher is critical of unconventional perspectives. Rather than trying to stand out, put your energy and focus into creating an argument that is clear and can be substantiated.

You may be wondering why I have dedicated so much time to the thesis, which is literally only a couple of sentences in the introduction. The reason is that in any essay the development of a solid thesis falls into the most valuable 20% of effort. Your thesis can make or break your essay and is therefore instrumental in winning marks.

The next component of an **introduction** is a brief mention of **context**, situating your argument in its relevant time, place, circumstances, or value system. This can be 1–2 sentences.

The next element to follow in an **introduction** is a brief **overview of the main points** you will make in each body paragraph.

The general guidelines for this component are as follows:

- There will usually be approximately 3–4 main points outlined in the introduction, corresponding directly with the number of paragraphs in the essay – e.g. if you make three main points, you will ideally have three paragraphs.
- Each point in the introduction should be brief, with 1–2 sentences conveying the main argument of each paragraph, and these will become the foundation for your 'topic sentences' at the start of each paragraph.
- Each point should support the central argument or insight in the thesis:
  - o If a point falls outside the scope of the thesis →expand the thesis or narrow the point.
  - o If a point contradicts the thesis

→re-evaluate the thesis or the point.
    - If a point does not relate directly to the thesis
      →refine or re-evaluate the thesis or the point.
- Each point needs to be distinct enough from other points
    - If there's too much overlap and repetition it won't work.
- Each point must be substantial enough that you can write a whole paragraph to support or prove it:
    - A good point in the introduction that can't be supported in the paragraph will compromise the essay.

To demonstrate this, let's go back to the *Twelfth Night* example. Bear in mind that the example does not require familiarity with the play, it simply illustrates the process in a tangible way, because it's hard to grasp hypotheticals. Recall the thesis:

'William Shakespeare uses the tragi-comedy form of *Twelfth Night* to create complex characters who evoke empathy as they suffer the consequences of their social transgressions. As the play shifts from its comedic exposition to its tragic end, Shakespeare warns of the harmful ramifications of defying social boundaries.'

Assuming this is a 3-paragraph essay, the three key points could be as follows:

- Olivia evokes empathy, suffering due to her transgressions in love
- Viola evokes empathy, suffering due to her transgression of gender boundaries
- Malvolio evokes empathy, suffering due to his transgression of social class.

You will notice that all three of these points, although distinct, relate directly to the binding thesis, as they are all characters who evoke empathy through suffering the consequences of their unique social transgressions.

TIP

Topic sentence 1 + Topic sentence 2 + Topic sentence 3 = THESIS.

A Year 9 English teacher provided a very helpful tip: think of a thesis statement as the sum of your topic sentences (or main points). This does not mean you should literally string your three topic sentences together to create a thesis. It means that instead of starting with a broad thesis and creating topic sentences for each paragraph, it can be more beneficial to **first create the three topic sentences** and then **reverse engineer an insightful thesis** which encapsulates the main ideas you plan to explore in the body of your essay. This helps avoid problematic thesis statements which are too broad or too narrow, or which sound impressive but can't be sustained or supported in the body of the essay.

Therefore, before launching into your introduction, it can be helpful to do a brief skeleton plan of your essay. Once you know what your main argument in each paragraph will be, and have crafted appropriate topic sentences, it will be easier to write a more sophisticated thesis leading to a cohesive introduction.

Note:

- If your essay has four paragraphs, your thesis will incorporate all four topic sentences.
- Remember the importance of ATQ – regardless of your approach, ensure you answer the question clearly and explicitly throughout your topic sentences and thesis.

***

The final part of an **introduction** is a brief **link back to the question** to close the loop and reinforce ATQ.

Putting all this together, the complete introduction for our Twelfth Night example might be as follows:

'William Shakespeare uses the tragi-comedy form of *Twelfth Night* to create complex characters who evoke empathy as they suffer the consequences of their social transgressions. As the play shifts from its comedic exposition to its tragic end, Shakespeare warns of the harmful ramifications of defying social boundaries. This is illuminated in Olivia's transgression of love expectations, Viola's transgression of gender boundaries, and Malvolio's transgression of social class. As these characters are transformed from comedic to tragic in their suffering, Shakespeare evokes empathy in audiences to warn of the repercussions of crossing social boundaries.'

Note: There are many other ways to interpret the question and construct or word this thesis and introduction; this is just one possibility to demonstrate the process and considerations. The main thing is ensuring you explicitly ATQ, provide an overarching thesis, contextualise and headline the main point you will make in each paragraph, and link back to the question.

TIP

The introduction can be a game changer!

Marks can be won or lost in the introduction. Most people assume that the introduction is merely a structural formality and underestimate the impact of a strong – or weak – introduction. In fact, the introduction is one of the most significant determinants of an essay grade because it's the 'preview' of the essay and will be judged by the marker.

A strong introduction which directly answers the question (and addresses the directive term), posits an insightful thesis, and signposts strong paragraph arguments, will signal to the marker early on that a student knows their content AND knows how to apply it. This primes the marker positively from the start. The rest of the essay certainly needs to deliver, but it's helpful when a marker has a favourable impression of your work at the outset and understands exactly where your argument is heading. This can win marks.

The opposite is unfortunately also true. If the introduction is sloppy, confusing, doesn't answer the question adequately, or fails to create a clear guide for what follows, the marker may be negatively primed and this can compromise marks. Of course, strong body paragraphs and clarity of argument evolving later in the piece can be redeeming, but the marker invariably has to work harder to spot opportunities to award marks if the essay is not set up well in the introduction.

Bear in mind that when markers have hundreds of essays to mark, often under time pressure, they don't necessarily have time to figure out a confusing argument, even if it becomes clear later. This is why a strong, clear introduction helps markers identify quickly that boxes have been ticked, allowing them to award marks easily. As we have seen with the marking criteria, there is an element of subjectivity and relativity, so if the introduction is strong, it can generate great marks, and conversely, if it's weak, this can create an obstacle to achieving top marks.

Part 2 – The body paragraphs

The body paragraphs form the 'meat' of the essay. This is where the arguments are progressed, evidence is presented, and the thesis is proven. The bulk of the writing in an essay will be devoted to the body paragraphs. While the introduction is very important to set you up for success, the paragraphs must be executed well to translate to marks.

> **TIP**
>
> **Know your body paragraph structures well – there is an expected format.**

Body paragraphs tend to follow a specific structure, which may vary slightly by subject.

Let's first look at a conventional history paragraph structure. I am using history as an example because it's currently compulsory for Years 7–10, and has a strong essay emphasis, meaning you will certainly be required to apply this learning from the start of high school. If you can master these structures and conventions early, it creates an academic advantage.

A **history** body paragraph will tend to have the following components:

- **T**opic sentence/**P**oint – states the key point/argument in the paragraph and should:
  - o Demonstrate how the paragraph will ATQ, ideally including key words from the question
  - o Align with the outline of the relevant point in the introduction
  - o Directly support the overall thesis
  - o Link with the previous paragraph for sustained flow.
- **E**vidence supporting the key point in your topic sentence:
  - o This will be contextually specific, relevant historical evidence.
- **E**xplanation or analysis of evidence:
  - o Demonstrates how/why this supports your key point/thesis.
- Depending on the length of the essay and the mark allocation, you will usually have 3–4 pieces of Evidence and Explanation per paragraph.
- **L**ink back to the question at the end of the paragraph, tying it all together.
- A full essay will usually have 3–4 paragraphs.
- Ensure your paragraphs are balanced with similar length and density of evidence:
  - o Avoid a situation where one paragraph is 400 words with five pieces of evidence, and another is 200 words with two pieces of evidence.

- Ensure your paragraphs flow to create a cohesive whole:
    - o Avoid creating disconnected paragraphs which fail to build towards a cohesive argument.
- You might hear teachers refer to 'TEEL' or 'PEEL' paragraph structure, which is a handy acronym for the components:
    - o TEEL = topic sentence, evidence, explanation, link
    - o PEEL = point, evidence, explanation, link.
- This structure is established in Year 7 and continues to be used as a foundation for more complex work all the way to the end of Year 12.
    - o You will be expected to provide a greater amount and depth of evidence and explanation as you advance to the senior years, so a typical paragraph would contain P, EE, EE, EE, L.

Now let's look more closely at **English**:

You will notice that I keep singling out English! This is intentional because English can be tricky, and since it's compulsory all the way to HSC it will be relevant for ALL students for ALL year groups. It's therefore worth spending time understanding the English essay and paragraph structure, given that every single student will need this structure throughout high school.

An **English** body paragraph will have the following components:

- **T**opic sentence/**P**oint – states the key point/argument in the paragraph and should:
    - o Demonstrate how the paragraph will ATQ – preferably including key words from the question and/or prompt, making a clear link
    - o Align with the outline of the relevant point in the introduction
    - o Directly support the overall thesis and key conceptual insight
    - o Link with the previous paragraph for sustained flow.

- Evidence (or Example) – supporting/proving the key point in your topic sentence:
    - o This will usually be a quote or textual reference from your prescribed text
    - o Or it will be a scene in a film analysis
    - o Evidence only has value if it builds towards **proving** the key point and ATQ:
        - o A great quote adds nothing if it's not directly supporting your argument – irrelevant evidence does not win marks.
    - o Evidence in each paragraph is usually presented in chronological order as it appears in the text.
    - o It's preferable to include evidence from the beginning, middle, and end of the text to create a holistic argument – if evidence is concentrated on one portion of the set work, this can cost you marks.
    - o Try to embed quotes and evidence naturally into sentences to improve flow.
    - o The evidence MUST contain a **technique**....
- Technique – this is the literary or film technique contained within the evidence:
    - o The technique only has value if you can demonstrate **how** it supports or proves the key argument and thesis.
    - o Naming a technique will not win marks if it doesn't prove your point.
- Analysis – this is **analysis** of **how** the **evidence AND technique** convey **meaning**:
    - o This must show how **meaning** is created, supporting the key point and thesis.
    - o Analysis that does not demonstrate **meaning** and support your argument directly will not score marks.
    - o TIP As mentioned earlier in the section on directive terms, try to include linking words into your sentences, such as 'conveys', 'reveals', 'reflects', etc. to prompt your exploration of meaning.

  - o TIP Intersperse key words from the question or prompt, creating explicit links for the marker and keeping yourself on track to ATQ throughout.
  - o TIP Analysis for each piece of evidence does not need to be long – if you can deliver meaningful insight in a clear and succinct way, this elevates sophistication and is preferable to lengthy or repetitive rambling.
- You will usually have 3–5 pieces of evidence (with technique and analysis) per paragraph (depending on the length of the essay and the mark allocation).
- **L**ink back to the question and thesis at the end of the paragraph, tying it all together.
- Depending on the length of the essay and the mark allocation, you will usually have 3–4 body paragraphs.
- Paragraphs may be structured by character, theme, or literary progression – this will be guided by the unit you are studying and your central argument.
- Ensure your paragraphs are balanced with similar length and density of evidence:
  - o Again, avoid a situation where one paragraph is 400 words with five pieces of evidence, and another is 200 words with two pieces of evidence.
- Ensure your paragraphs flow to create a cohesive whole:
  - o Avoid creating disconnected paragraphs which fail to build towards a cohesive argument.
- As previously noted, teachers may refer to the 'PEEL' structure (discussed above in 'history'), but more often in English teachers propose the 'PETAL' structure, which more appropriately relates to English. The acronym represents:
  - o PETAL = Point, Evidence, Technique, Analysis, Link.
  - o This structure applies from Year 7 to the end of Year 12, and a full paragraph will typically contain P, ETA, ETA, ETA, L.
- In higher years, you will also be required to include explicit **context** statements, which are usually placed just after the topic

sentences, although a sophisticated response will often have context woven through the paragraphs as appropriate…

- o Context statements are necessary, because it's only possible to fully analyse a text by considering relevant influences which impact the meaning of the text or the purpose of the composer. This is especially crucial when comparing texts written in different contexts or with differing values.
- o Context can include values or conventions of the era, or the personal context of the composer.
- o Context can be social, political, cultural, environmental, literary, personal (or anything else that may have influenced the work).

Part 3 – The conclusion

By this stage you have hopefully written an amazing introduction and fabulous body paragraphs, and you likely just want to get the essay finished, but skimping on the conclusion, or forgetting to write one can let you down at the last hurdle.

The conclusion does not need to be long (it should be shorter than the introduction), but must have the following components:

- Restate your thesis and ATQ
- Briefly recap your key point from each paragraph
- Ideally add a final comment linking back to the question and closing the loop
- Small TIP: don't start your conclusion with words like 'in conclusion' or 'in summary' – teachers don't seem to like that ☺
- Do not introduce any new points, arguments, or evidence in the conclusion.

Question:

If you run out of time, should you favour the last body paragraph or the conclusion?

Essay exams are notoriously tight for time. In the HSC you will only get 40–45 minutes for each essay, including planning time. That's not much time to write a 1,000+ word essay, so it's not uncommon to run out of time having to make a tough decision.

Imagine a scenario where you are part way through your final body paragraph in an English essay, having only analysed two of the four pieces of evidence, and you realise you only have three minutes left. In this situation, you know something will have to be sacrificed. Do you finish your analytical paragraph and cut out the conclusion? Or do you abandon the paragraph and write the conclusion?

This is not a great situation, and it would be ideal to avoid it, but if it happens (and we have seen it happen), there are two schools of thought:

- Some teachers advise students to leave a few lines of space and move on to write a conclusion, even if it's only brief, and then use any remaining time to go back and write as much as possible to add value to the final paragraph. Their reasoning is that the conclusion completes the structure as it wraps everything up, summarises the key points and restates the answer to the question. Without a conclusion the essay could be considered incomplete.
- Conversely, other teachers express the alternative view that students should complete the analytical paragraph at the expense of the conclusion, since marks are awarded for answering the question, and the question is primarily answered in the paragraphs through the evidence and analysis. This view considers that if the analysis is abandoned, the incomplete paragraph may cost marks because it might not have fully answered the question.

This is a tricky scenario, and conflicting viewpoints create a challenge. My advice would be to do a quick assessment of how far you developed your argument in the final body paragraph. If you have managed to mostly answer the question and support your point with the evidence and analysis you've already done, then I would move onto the conclusion. If you have not yet answered the question in your final body paragraph, then do this as efficiently as possible, leaving enough time to do at least one concluding sentence, summing up your overall answer to the question. Given the lack of consensus, I would advise students to find out the preferred approach of their teacher or marker.

***

Essay and paragraph structures, such as those presented here, probably seem overwhelming at first, but they provide a valuable scaffold, forming the basis of most extended responses in Years 7–12. These structures, along with their useful acronyms, can be viewed as a formula which, if correctly applied, will automatically guide you towards the creation of a response incorporating many of the structural and qualitative marking criteria necessary for top grades. The requirements also don't change much as you progress through high school, so practising this skill from Year 7 means you will accumulate tremendous gains in mastery by the time you get to Year 12, developing a refined ability to give markers what they want. Once you are fluent in applying the formula, it frees up cognitive space, allowing you to focus your efforts on building and refining the sophistication of your responses.

Note: I have only addressed English and history structures here, as they are broadly applicable, benefiting all students. However, there will be variations for other subjects, and my advice is that students ensure they become familiar with the structures and scaffolds for other subjects as soon as they become relevant.

We have now spent significant time delving into structural and content requirements, but there are a few common-sense macro considerations which we can't ignore when it comes to giving markers what they want.

Will the marker be able to understand what you have written?

We have covered a lot of ground related to giving markers what they want, but there are a few more fundamentals to bear in mind. You could compose a strong, sustained, well-structured response that answers the question and successfully addresses the marking criteria, but if you forget some of the basics, it could result in disappointing marks.

TIP

Do recognise the value of clear and simple language.
Don't use overly complicated vocabulary just for the sake of it.

Marking rubrics often allude to 'sophistication' in the higher mark ranges, and while a sophisticated argument is rewarded, it's not an invitation to fill your response with inaccessible vocabulary from the thesaurus. I say this only half-jokingly, because I have seen students make this mistake. In trying to elevate their language they look up elaborate synonyms and ultimately their writing appears inauthentic, disjointed, and may not even make sense.

It's important to recognise that sophistication comes from **clarity of insightful ideas**, rather than the utilisation of sophisticated words. A highly sophisticated and conceptually nuanced argument can be skilfully conveyed in plain, simple language. A clear argument, which is easy to follow, will always score better than a convoluted argument packed with impressive words. Sometimes less is more. My advice is therefore to ensure your argument is clear, the sophistication is demonstrated in your ideas and expression, and your response is easy to read and follow. This is the best strategy for winning marks, as clarity makes it easier for the marker to tick the boxes.

TIP

Avoid long, rambling sentences.

To enhance the clarity of writing, avoid very long sentences. If a sentence is getting too long, rather break it into shorter, more manageable sentences.

TIP

Watch your handwriting – if they can't read it, they can't mark it!

This applies to **all** subjects and year groups, whether it's an English essay, a chemistry question, or a maths calculation.

Teachers will remind students about handwriting throughout their time at school, and some students may dismiss this with comments like "the teacher hasn't complained about my handwriting" or "my classmates have worse writing and they are doing fine". If you receive this feedback from teachers, I advise you to take it seriously and attend to it as soon as possible. The longer you ignore poor handwriting, the worse it gets. This becomes increasingly problematic in the senior years as exams require greater volume of writing under increased time pressure. If you have not sorted this by Year 11, it's almost too late. This may seem like a minor issue, but I have personally known a student who was marked down in a Year 12 essay exam because the marker could not decipher the handwriting. This not only cost them valuable marks, and compromised their rank going into HSC finals, but unfortunately also cost them a school 'top in subject' award. The harsh reality is that teachers and HSC markers (even the nicest ones), don't have time to decipher illegible handwriting.

TIP

Always proofread your work.

Whether it's an essay, a research project, a practical report or a homework assignment – always proofread your work to correct typos, spelling mistakes, grammar, and punctuation. Work that is handed in with errors will be negatively perceived and may cost you marks.

A final question: have you made the marker's job easy enough?

TIP

Markers want to award marks … so make it as EASY AS POSSIBLE for them!

Markers want students to do well. They prefer to give marks rather than withhold them. Your job is to make this as easy as possible, creating clear opportunities for markers to be generous. The harder a marker must work to **find** places to award marks, the lower your marks are likely to be. This applies to absolutely everything you do, whether it's a short answer, an essay, or a calculation.

***

All the strategies and tips covered so far in this section have been geared to help you make things clearer and easier for the markers … to give them what they want … so they can give you what you want, which is top marks. Whether it's answering the question, having legible handwriting, ticking boxes on the marking criteria, providing the appropriate structure, writing in a clear and simple way to make a strong argument, or ensuring you have a thesis with an overarching insight … these skills can be learnt, practised and refined over time, accumulating to create your blueprint for excellence.

***

In this chapter, we have looked closely at the demands of assessments, and gone into detail on how to decode what is being asked and tested, what you need to do, and how to answer questions in a way that gives markers what they want, but what comes next?

CHAPTER 8:
Assessment preparation and exam technique

Assume you have been issued an assessment notification. Based on everything we have covered so far, you hopefully know what you are being asked to do, what content you need to know, you are confident that you will be able to decode and answer the question(s), and you know how to execute the task to give the marker what they want. But to give yourself the best chance of success there is valuable preparation to do and exam techniques to master. A bright student who is unprepared and lacks technique will struggle to do well. The right amount and quality of preparation, along with skill in exam technique, bridges the gap between knowing what to do and executing it to a high standard, facilitating your achievement of academic excellence.

You may be confused as to why we are backtracking to talk about preparation and exam techniques, after having already discussed how to answer questions and address marking criteria. The reason I am presenting in this order is because it's only possible to talk about preparation if you have a solid grasp on the scope of what you're preparing **for**, and it's only possible to finetune exam technique if you are confident in all other aspects. If you have goal clarity from the start, the preparation and finetuning of study and exam techniques to get you there become more logical and meaningful.

Once again, there are certain types of preparation common to most subjects, years, and assessments, but there are some activities which are specifically tailored to certain scenarios. Let's start with some of the broadly applicable types of preparation.

1. Making notes

Notes are the foundation of all study. For all subjects. For all year groups.

Making notes is therefore crucial for academic success, and you would be hard-pressed to find a high-achieving student who does not diligently make notes. I mentioned this in an earlier chapter, but I will restate it here – making notes is a Quadrant 2 activity which means it's usually not urgent (unless your assessment is very soon), but it's extremely important, leading to higher-quality preparation and outcomes. Making notes is part of the most valuable 20% of work, as great notes generate a fantastic return on your effort.

I am sure I don't need to sell this point, since making notes is a well-established study technique, but here are some of the key benefits:

- Notes are your primary study resource – without notes, what do you study?
- It's easier and more efficient to study from your own notes rather than directly from textbooks or other sources which sometimes have superfluous information or cover content in a different order.
- Creating a single set of notes combining class notes, textbook notes, worksheets, etc. makes it easier to study, rather than jumping between various sources.
- Making notes clarifies and consolidates understanding of subject content.
- Making notes highlights gaps in knowledge.
- It's easier to remember content that is in your own words and format.

- The process of making notes helps commit the information to long-term memory, meaning that making the notes is a valuable form of revision.
- Making notes helps you feel organised and in control.

There are many approaches to making notes, depending on individual preference, but I will provide some **best practice guidelines**:

- **Handwritten** notes are usually best – based on science and personal experience.
  - The act of handwriting, which employs fine motor skills, engages various areas of the brain, leading to better comprehension, memory, and learning.
  - The more you handwrite, the better and faster your writing will be, and given that the HSC is currently still handwritten, this consistent writing practice helps build essential skills over time. This is especially important given the increased reliance on technology both at school and home.
- However, there may be occasions when **typing** notes is preferable or necessary.
  - Ideals should be balanced against time constraints and typing is faster.
  - There may be scenarios in which typing is more efficient, for example, in Year 12 for content heavy subjects such as biology, history, business and legal studies, etc. where notes involve hundreds of pages and may need to be supplemented or tweaked retrospectively over time.
  - But still – I would encourage handwritten notes wherever possible.
- Put content into your **own words** – whether you are handwriting or typing notes.
  - This is based in science, since reconfiguring, summarising, or rephrasing information creates deeper engagement with the content, assisting with comprehension and retention.

- Transform the content from words and sentences into **visuals** for greater learning.
  - E.g. making mind maps, flow charts, flashcards, or diagrams similarly activates deeper engagement, facilitating better learning and retention.
  - Visual representations also make content easier to remember and study, because they tend to be less dense.
- The type of notes you make should accommodate your personal **learning style** – for example, some people prefer to learn using words, while others prefer diagrams.
- Use **colour** and highlighters to make key pieces of information stand out.
- It's not advisable to copy/paste notes:
  - This requires minimal brain activation, limiting process-driven learning.
- It's also not advisable to use notes made by another person:
  - This may be tempting to save time, but is completely passive and may limit your learning, especially if they are written or constructed in a way which doesn't suit your preferred style.

There are also a few **traps** to be aware of when making notes:

- Balance **function** and **aesthetics**:
  - Don't spend too much time making notes perfect or pretty – notes fulfil a function and do not need to be a work of art.
  - The danger of getting carried away with making notes beautiful is that it consumes valuable time, with diminishing returns.
  - You need to recognise when the notes are good enough functionally to generate your desired results – and then stop!
- There is an art to knowing **what** to make notes on:
  - Writing down absolutely everything on a topic or rewriting the textbook is counterproductive. This wastes time and makes it harder to study later, because with too much content it can be difficult to remember everything, and harder to sift out what is most important.

    - o On the flip side, if your notes are too vague or brief, you may not have enough depth of information to study from, and this can compromise your ability to provide quality answers at the appropriate level.
    - o There is no formula for exactly what or how much to write, because it depends on the subject, the year, your learning style, and personal preferences. This entails a bit of trial and error until you figure out what works best for you. The best advice is to start figuring this out in Year 7, so you can build on these skills as you progress through high school.
- Ensure your notes are **clear and legible** – there is no point making notes that you won't be able to understand later.

TIP

Take notes in class.

As a rule, the notes you take in class will be high value. Teachers focus their teaching on the most important aspects of the curriculum, and they generally test what they have personally taught … so you want to write it down.

Teachers have varying teaching styles – some may spend class time talking and explaining, others may teach from prepared slides, some may prepare student booklets, and some may teach from worksheets. Whatever the teaching style, it's important to make notes in class that will later become the foundation for your study.

A note of caution

Be careful to balance making notes with listening and understanding what is being taught. There is no point writing down everything the teacher is saying if you can't also focus on conceptual understanding. Having comprehensive notes that you don't understand later is not helpful, in fact it can be harmful. So, first and foremost, listen, then make notes on important things you will need to remember later. If you have paid attention in class and taken decent notes, that's half the battle won!

TIP

It's ok to make class notes in shorthand – they do not need to be flawless.

Your class notes don't need to be grammatically correct or written in full sentences. Using shorthand, abbreviations, key words or rough diagrams will be sufficient. If you miss something it's ok, the notes can be fleshed out later, but this is where the 80/20 principle comes into play – allowing you to make notes which are 'good enough' as a basis for revision, while balancing listening, comprehension, and keeping up.

TIP

The best notes are a consolidation of class notes plus textbook notes.

I still stand by my assertion that class notes are very important, but class notes alone will not suffice. Although teachers hopefully teach the most important content, this will usually need to be supplemented by the textbook, which is designed to cover the entire curriculum. Even with the best teacher, there may still be gaps in knowledge if students rely solely on classwork. Some of the possible reasons for this are:

- You may miss something important when taking notes in class
- The teacher may omit certain details (this can happen for various reasons)
- Teachers are also applying the 80/20 principle to their lessons
- You may be absent and miss important content, and so on ....

In the event of a weak teacher, or one who doesn't communicate clearly, or doesn't resonate with your particular learning style, the textbook becomes especially crucial as it might become your primary study resource. My aim is not to be controversial or negative, but the reality is that even at the best schools, not all teachers are equal. Through high school we have unfortunately been faced with this reality on a couple of occasions (thankfully only a minority) and without the textbook, learning would have been significantly compromised. Students who utilise the textbook resources properly to create comprehensive notes are better equipped when it comes to assessments. It's important for students to be proactive and self-sufficient. Those who take ownership of their learning are more likely to succeed academically.

So, although class notes are a valuable foundation for your study notes, in almost all cases you will need to refer to the textbook and supplement. The best and most comprehensive notes are therefore a hybrid of class notes and textbook notes.

TIP

Make notes ASAP after learning the content.

The best approach to notes is a proactive one, where you consistently keep up to date with your learning, making notes while the content is fresh in your mind. The absolute ideal is to spend some time each day after school making notes based on what you learnt during the day… but honestly as I am typing this, I already know from experience that although this is the best approach, it's also unrealistic for most students. Afternoons tend to be taken up with sport, activities, homework, social and family commitments, appointments, etc. and finding time to make notes proactively for the 4–7 subjects you had during the day is near impossible.

So, if we approach this more realistically, then perhaps it's worth planning in some time on the weekends to catch up on notes for the week. This is more feasible and students adopting this approach will be in a very strong position when assessments roll around.

CAUTIONARY TIP

Don't neglect the notes or leave it too late!

In reality, many students won't proactively make notes and will only start once they receive the assessment notification. This is less ideal, but it's still ok. If you start making notes as soon as you get the notification, you should still have time to do a quality job, especially in Years 7–10. It becomes riskier in Years 11–12. The danger of not prioritising notes and leaving them to the last minute is that they move into Quadrant 1, which is where important tasks suddenly become urgent. This is stressful and can compromise the quality of your notes. Even if you manage to create a great set of notes, if you leave it too late, you won't have sufficient time to study them. Either way, it's not likely to help you achieve the best results.

The other consideration is how much other work you need to juggle. If you have an exam block and receive all your assessment notifications at once, it will be a challenge to try and make all your notes for all your assessments in the time available.

TIP

Structure notes according to the 'syllabus dot points' and 'inquiry questions' if possible.

Syllabus dot points and inquiry questions are terms you will encounter frequently. These outline the skills, knowledge, and understanding that students are expected to acquire through engagement with the curriculum. Teachers usually provide these for each unit of study and then teach specifically to these outcomes. It's therefore helpful to use these as a structural guide for your notes. Following the list also helps you identify any gaps which need to be filled. This is especially relevant in content heavy subjects such as the sciences or history. The curriculum information is also available on the NESA website.

TIP

Make your notes on loose-leaf paper and put them in a file at home.

Some students make notes in their classwork book, but this can make things tricky. It's hard to separate class 'notes' from class 'work' when trying to study, and it's difficult to integrate supplementary notes in an effective and cohesive way.

If you make your notes on loose-leaf paper you can easily add to them or insert pages, and you can file all your notes in one place in an organised way, making it easier to study when the time comes. If notes are typed, it's good to print and file them regularly.

TIPS for English notes

Annotate and highlight IN your set work wherever possible (as you study it in class).

For English studies centred around written set works such as poems, plays, or novels, the most efficient way to make notes is highlighting and annotating in the text itself. For example, if the teacher is reading through a poem or play, pointing out important quotes and literary techniques, then highlight the quote and write your annotations or notes next to it. Some people are averse to writing in books – I used to be one of those people – but it's a lot easier and more time efficient to study for an exam on your Shakespeare play if it's been annotated within the text, rather than referring to a separate set of notes.

Additional TIP for English
For a film set work, make a note of timestamps.

If the set work is a film – write down the timestamp for key scenes, along with notes on techniques and analysis. There is nothing more frustrating and time consuming than searching through the film to locate a scene later. If you have the relevant timestamps, this will save you a lot of time when studying or preparing analysis.

Additional TIP for English
Make a 'quote bank' or 'scene bank' for each set work.

You may not yet be familiar with the term 'quote bank'. Essentially, a quote bank is a list of key quotes, accompanied by their techniques and associated analysis. This is a valuable part of your English notes. There are different ways to organise and present this information, but we have found that summarising it in a table format organised by theme is simple, clear and easy to revise.

Here is an example taken from a quote bank prepared by my daughter for her study of Charles Dickens' *Great Expectations* (Mitchell, 2003). Note: familiarity with the novel is not required, and this is for illustration purposes only.

Table 8.1 A sample quote bank

Theme	Quote	Technique	Analysis
Corruption	"I soon contracted expensive habits"	Metaphor of disease	Conveying the corruptive nature of wealth

A 'scene bank' would be the film equivalent. Here is an example from my daughter's scene bank for her Year 12 study of Jane Campion's film *Bright Star* (2009). (Again, familiarity with the film is not required.)

Table 8.2 A sample scene bank

Theme	Scene and Timestamp	Technique	Analysis
Immortalisation of art	Ballroom scene (timestamp – 10min)	Framing using the visual motif of the mirror	Imitates a framed artwork, conveying appreciation of moments eternally frozen in time

These quote banks or scene banks can get quite long, but if they are done properly, they are a very valuable resource for studying and preparing analysis.

Note: I have used Year 12 examples, but this same template is both suitable and recommended for English from Year 7.

The most efficient way to construct a quote or scene bank is to create the structure early, populating it as you move through the study of the text. This is easier and quicker than trying to go back and create a whole quote bank from memory later. It also helps to keep the evidence in chronological order.

Cautionary note

Many students construct quote banks by simply compiling a list of quotes. But without the accompanying techniques and analysis, a list of quotes is almost useless. Students who memorise quotes in isolation are disadvantaged in an exam if they haven't also prepared and studied the techniques and analysis, because it can be very difficult to make this up on the spot under time pressure, and if not done properly, this will likely compromise the quality of argument and ability to answer the question.

<u>TIPS</u> for other English notes

Make notes on the context, values, purpose, key message, and form of the set works.

In addition to the quote banks and annotations in the set works, you will also need notes on the context and values that influenced the composer of the text or film, along with the form, key purpose, and key message of the text. In Year 7 or 8 this information may not be strictly required, although including it would be rewarded. But you will want to revisit this tip later, because certainly from Year 10 onwards you will be marked down if you don't include these aspects in your argument. These essential elements tend to be forgotten, but if you have brief notes that you can revise before an assessment, it will be much easier to incorporate into your writing.

<u>TIPS</u> for maths notes

Make notes for maths too!

Maths is not a subject one naturally associates with notes, but they are very beneficial. Having said that, mathematics notes are different to the type of notes you would make for theory-based subjects. Maths notes can be brief and will often include:

- Brief theory or conceptual explanations
- Key rules
- Formulae
- Worked examples
- Difficult questions
- Alternative methods

Although concepts must be clearly understood and some formulae require rote learning, maths notes are usually not made for memorising or studying, but rather for reference and reminders. This is because maths is not a subject requiring extensive study and rather requires **extensive practice**.

Maths notes should not take very long to prepare, and this is one subject where you probably don't need to make notes after every lesson. A good approach is to make brief notes from the textbook each time you complete a topic or chapter.

***

Now, assume you have comprehensive notes for your various subjects. What next?

2. Studying and revision

I'm sure you can anticipate that the next step after making notes is studying and revising them. While notes are necessary for working smart, constituting part of the crucial 20% of effort, what you do with those notes will ultimately determine your academic success. Both components are therefore necessary to create true academic value. In other words, poor notes will likely lead to poorer outcomes, no matter how hard you study, and great notes will contribute very little if you don't make optimal use of them. You therefore need BOTH great notes and effective study to achieve great grades.

Poor notes + study = poor grades ✗
Great notes + ineffective study = poor grades ✗
Great notes + effective study = great grades ✓

We have already addressed how to make good notes, so now we turn our attention to the 'study' part of the equation. There are a variety of methods that can be used for study and revision, and depending on individual learning styles certain methods may work better for different students. Some methods are also more suitable for specific subjects or types of assessments. Before delving into the specifics, let's take a macro view.

The main purpose of studying and revising your notes is:

- To develop a thorough **understanding** of the content
- To develop familiarity with the content so you can **remember** it
- To become comfortable with the content so you can confidently **apply** it when answering assessment and exam questions.

You can see there is a progression which must take place for effective study to occur. To simplify let's represent this as an equation:

Understand + Remember + Apply = **SUCCESS**

As an equation, if any of these variables fall away, the resulting success will diminish.

Put another way:

- If you don't understand it, you won't be able to ATQ
- If you understand it but can't remember it, you won't be able to ATQ
- If you understand and remember it, but can't apply it, ATQ will be limited.

For effective study to occur you need to **understand**, **remember** and **apply** the information. Let's look at these one at a time:

<u>How to ensure you fully understand the information:</u>

Most of the understanding has hopefully occurred prior to the revision stage as you have listened in class, asked questions, consulted the textbook, and made comprehensive notes, but this is not the end and there is more to be done …

- Highlight and review your notes – this consolidates understanding because:
  - You may be rusty on content you did further back
  - You may spot gaps in understanding
  - With time you may have enhanced perspective
  - Revising a full note set improves holistic understanding
- Teach or explain the content to anyone who will listen – a friend, a family member, a pet, or if nobody wants to hear it, explain it to yourself out loud:
  - Even if you think you understand something, when you explain or teach it, it can throw up questions you didn't know you had, and at the very least it helps to clarify your thinking and cement your understanding.
- If something is unclear or you still have questions, ask someone.

<u>How to ensure you remember the information:</u>

To **remember** information, it needs to be committed to **long-term memory**.

I thought it might be helpful to briefly reference the work of Hermann Ebbinghaus, a German psychologist who pioneered the scientific study of human memory and the psychological concept of the 'forgetting curve'. In short, the forgetting curve demonstrates that memory retention drops rapidly after learning if material isn't reviewed regularly (Ebbinghaus, 1913). The implication of his work is that spaced repetition slows the process of forgetting, thereby enhancing memory. Therefore, students wanting to employ effective study routines should revise their work at spaced intervals to optimise information retention.

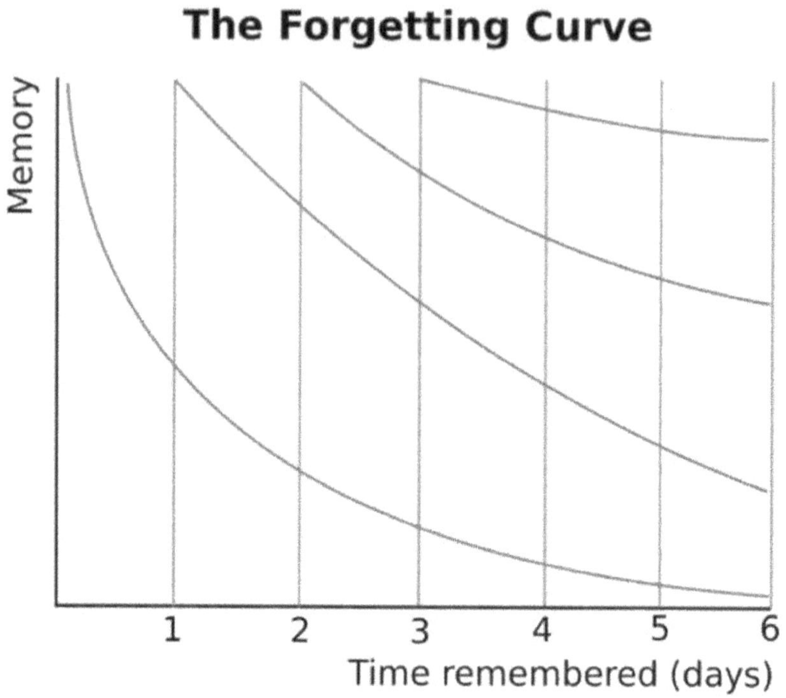

Figure 8.1 Ebbinghaus forgetting curve showing how memory retention declines over time without review. Based on a public domain representation from Wikimedia Commons (ForgettingCurve.svg)

This simple graph shows that your initial learning will be almost completely forgotten within a few days if it isn't reviewed. However, revising your notes regularly slows down the forgetting process, helping to shift the information to long-term memory.

> **TIP**
>
> **Review early. Review briefly. Review often.**

Multiple short, repeated study sessions are more effective at building memory than one long cram session. From a practical perspective you could optimise retention by reviewing material within 24 hours of learning it, revising again within 3–4 days, and then again the following week.

- These are a few revision techniques that could be used:
  - Repeated revision – going through your notes multiple times.
  - Reading your notes out loud – the process of seeing, speaking, and hearing engages multiple senses, enhancing learning and memory.
  - Making summaries of your summaries and transforming information into diagrams, flowcharts, mind maps, or flashcards – this promotes active recall.
  - Testing yourself by looking at a heading and trying to see how much you can remember without looking at your notes.
  - Asking someone to test you verbally on your notes.
  - A combination of techniques will be most beneficial.
- Active revision is always better than passively reading notes – the more you engage with the content in an active way (testing yourself, reading out loud, rewriting, or transforming the information), the better you will remember it.
- In our experience the best time to review notes is first thing in the morning when your brain is fresh, and again before you go to bed, as your brain processes and organises information while you sleep, ensuring it's filed in your memory for efficient retrieval. This is another reason why sleep is so vital – without sleep, the filing system is disrupted.
- Start your revision as far in advance of exams as possible – the more revision you do, and the more it is spaced out, the better you are likely to perform in exams.

A note and a couple of cautions

- For hand-in assessments, where you can freely access and reference your notes, there is less emphasis on remembering everything, but there is a caution attached to this, because even if interim assessments don't require everything to be committed to memory, there is a chance that subsequent tasks, certainly yearly exams, will require this step, so it's best to be proactive with studying your notes as you go. In the senior stages of high school more of your assessments will be in-class and exam style, so it's advisable to build these skills from Year 7, so when you need to activate this part of learning, you already have the techniques well practised and established.

- For an open-book assessment or exam, which sometimes happens in earlier years, you will be able to take textbooks or notes into your written exam to reference. This takes some pressure off studying and memorising content, but it can also be a trap. Some students will go into the exam under-prepared because they know they will be able to look up information, but flipping through pages of notes to find answers can disrupt focus and create additional time pressure, which can negatively impact performance. Additionally, the questions will seldom require simple regurgitation of content, and students who rely on looking for answers in their notes may not be able to adapt or apply the information to appropriately answer the question. Conversely, students who study for open-book exams as if they were closed book, generally perform better, because they have superior understanding and recall, and can devote their exam time to applying their knowledge to provide high-quality answers. My advice is therefore to treat open-book assessments as if they were closed book from a preparation perspective. This also lessens the load later when you will invariably be examined on the information in a closed book setting.

<u>How to ensure you can apply the information:</u>

Understanding and remembering are crucial for study success, but this isn't enough to win marks. We have established that you are unlikely to get questions asking you to simply regurgitate notes. Most questions you encounter, except very basic ones, require some degree of application, meaning you must be able to select relevant information from memory and **adapt** it to the specific demands of the question. This applies to all assessment types, whether they are hand-in or in class. This also takes revision one step further, going beyond studying and revising notes, to doing revision **practice**.

We spent a lot of time in previous sections learning how to decode and answer questions, but how do we build this application-based skill in the study phase? It's pretty simple ...

> **TIP**
>
> **PRACTISE! PRACTISE! PRACTISE!**

The only way to prepare for the application part is to consider the information from as many different angles as possible. This means you need to think beyond and around the theory. The best way to do this is to **practise answering questions**. The more questions you are exposed to, the better you will become at **adapting** and **applying** your knowledge in a way that will score marks.

My daughter's very wise Year 12 Extension 2 Maths teacher told the class that rather than 'thinking outside the box' when solving complex maths problems, they should aim to 'make the box bigger'! I love this idea, and it corroborates the importance of increasing your exposure through practice. It can also be applied to any subject.

The benefits of doing practice questions for any subject are therefore immeasurable:

- Thinking about Malcolm Gladwell's '10,000-Hour Rule' (2008), we know that skills build with accumulated practice ... more practice = greater mastery.
- You gain experience and familiarity with different question types – we have seen students who study hard and are completely comfortable with the content be thrown by how a question is asked, even though they have all the information needed to answer it. Remember that:

    - The same content can be examined in multiple ways
    - The same question can be asked in multiple ways
    - The more variety you have seen, the less likely you are to be thrown off by an unusual question, and the more likely you will be to confidently adapt.
- You expand your knowledge and skill by attempting questions of varying difficulty:
    - Doing a mix of easy, moderate, and hard questions builds confidence and skill, leading to improved performance as challenge promotes growth.

I can say with great confidence that practice is one of the most important strategies for academic success, especially when preparing for exam-style assessments. This is a Quadrant 2 task that, if properly planned, is usually not urgent and is extremely important for quality preparation, falling within the most valuable 20% of student effort. In fact, I would go so far as to say that no matter how smart or hardworking a student is, if they don't get enough practice, they are unlikely to achieve the grades they want.

Realistically, before you are in a position to practise the skills, you would already have to understand the content and ideally have studied it thoroughly. This is where many students falter, either because they underestimate the value of practice, or because they run out of time to get to this critical stage. This is why the skills we talked about in the early chapters, such as prioritising and time management remain crucial for success. We will talk more about this when we address assessment planning, but for now I wanted to impart the importance of making time to practise.

NOTE: This notion of practice applies to **all** subjects, whether it's practising mathematics problems, chemistry calculations, English essays, history source analysis, or anything else – practice is equally essential for every subject and every question type.

Knowing how important the practice component is, you may be wondering where to find the resources for practice questions. Well, they are literally everywhere!

Here are some suggestions (note that some are more relevant for certain subjects):

- Do the exercises in your textbook
- Do the chapter, topic, or module review questions in the textbook
- Do the questions in any activity books or workbooks that are supplied to you
- Complete worksheets, booklets, revision sheets, or questions given in class
- Ask your teachers for more – they have valuable resources to share
- Ask friends at other schools to share their resources (if they are allowed)
- Buy books with practice questions
- If all else fails, look online – there are endless free resources – just ensure that you only use reliable and reputable sites, because there is a lot of misinformation online
- AND DO PAST EXAM PAPERS!

This last point applies to **all subjects** and is so vital that I am giving it a separate heading!

3. Past exam papers

I am going to begin this important point with an anecdote. When my husband was in third-year medical school many years ago, he had various study techniques, including many that we have covered here. However, he always professed that the best and most valuable type of revision was doing past exam papers. This strategy served him extremely well. At the time he had a friend who was very smart and worked incredibly hard but was taking a different approach. This friend had made thorough notes and studied from not one, but two textbooks. He had accumulated extensive knowledge and felt confident going into the third-year final exam. Unfortunately, he failed that exam and consequently failed the year. The problem? He had devoted all his time to studying the theory and probably had more detailed knowledge than most other students in his cohort, yet this did not translate to marks in the exam. His biggest mistake was failing to practise his skills and, more specifically, neglecting to do the past exam papers. The consequence was that he had to repeat his third year of medical school. This friend learnt the tough but valuable lesson that past papers were a key requirement for success, and from that point on, he sailed through. This lesson has stuck with me over the years, and the principles are still equally relevant today, more than 30 years later.

Here are a few key takeaways from this story:

- While it's very important to make notes and study them, spending time making more notes from multiple textbooks adds less value, entering the area of diminishing returns, outside the most valuable 20%.
- Time spent on any task once it starts generating diminishing returns means time taken away from another key task. This carries a cost.
- If that time could be diverted to another task with a high return on effort, once again operating in the most valuable 20%, the overall return on effort would be higher.
- Studying theory is important, but doing practice questions and developing skills to apply the information is equally important.

When you are studying and preparing, particularly for unseen written exams (including the final HSC exams), there are multiple benefits to doing **past exam papers**:

- It creates **familiarity** with the format and structure of the exam, which increases your **confidence** in the actual exam.
- You are exposed to a **variety** of question types and formats which builds your **skill** base (e.g. multiple-choice, short answer, extended response, essays, calculations).
- You are exposed to a **cross section** of questions varying in length and difficulty.
- It allows you to **test** yourself **across topics**, rather than a hyperfocus on one topic at a time (which is often how revision is organised).
- You have the opportunity to practise **complex questions** which may combine multiple topics, concepts, or aspects of the syllabus. Classwork and revision questions are often organised in silos catering to a single topic, but for maths, examiners often create questions requiring the application of multiple concepts to arrive at a solution. Past papers are the best source of practice for such questions.
- It is the best way to put your theoretical **subject knowledge** to the test.
- It clearly shows **gaps** in learning which you can remedy before the real exam.
- It's the best way to build skills to **adapt and apply** your knowledge.
- It provides an opportunity to **improve exam technique and strategies**.
- It provides a valuable opportunity to navigate **mark allocations**, as you work out how to approach questions with varying marks attached – for example, you would approach a 2-mark question and a 9-mark question differently.
- It helps you develop crucial **time management** skills within an exam:
    - o To ensure you can complete the exam within the allowed time
    - o To ensure you allocate the correct proportion of time to various questions (often guided by the mark allocations).

- The added **bonus**: Often question types, or even exact questions, are repeated over the years. If students have worked through past papers, there is a reasonable chance they will encounter questions they have already seen and answered. This provides an obvious academic advantage as students will be able to answer the question more accurately and quickly than if they were seeing it for the first time. Even if the question is not identical, familiarity with the question type, wording, or format generates advantage through a practised approach.
- **Marking** past papers against model solutions and marking guidelines provides insight into how marks are awarded, providing a valuable opportunity to gauge where you may have lost or won marks so you can make the necessary adjustments in the exam.

Overall, **past papers are the best and most relevant source of revision and practice**, as they are the closest representation of what you will encounter in the exam. Past papers allow you to refine your strategies and iron out the kinks in advance, enhancing your exam performance when it counts. They are essential for academic success and students who neglect this aspect of preparation do so at their peril.

Furthermore, past exam papers are essential to working smart:

- They are a critically important Quadrant 2 activity (or, less optimally, Quadrant 1).
- They contribute to the most valuable 20% of effort with enormous academic impact, thereby helping to generate 80% of your results.
- They seldom take you into the area of diminishing returns, because additional practice expands your repertoire and builds expertise. As a guideline, in the earlier years you should do an absolute minimum of three past papers for each exam, with 4–5 being the norm. In the final years of high school, especially in Year 12, you

should do a minimum of five past papers per exam with no upper limit … the more you do, the better prepared you will be!

- This aligns with Gladwell's theory of excellence, in that practice accumulates to build mastery, which translates to results (2008).
- This also aligns with Dweck's **growth mindset** (2006) as students who take ownership of their learning recognise that efforts to practise, review, and refine their skills will build towards improved academic outcomes.

Where to access past papers:

- Schools usually supply past papers and model solutions prior to each assessment – it's advisable to do as many as you can. Ideally, all of them.
- If the school does not provide sufficient past papers, you can look on a website called *THSC Online* (https://thsconline.github.io/s/) which is a useful repository for papers from various schools – but be careful to only use papers which are relevant to the current curriculum.
- You can ask friends at other schools to share their past papers if they are allowed.
- For HSC – official past papers are accessible on the NESA website – these are essential for HSC exam preparation.

TIP

Revise your content quite thoroughly before attempting past papers.

To derive the greatest benefit from past papers, it's best to first revise and become familiar with the content. Attempting past papers when you haven't revised can be a demoralising waste of time.

If you have not developed a solid understanding of the material, or you can't remember the detail, trying to answer questions which require application of the knowledge is premature. You may find yourself unable to answer questions, and this can create unnecessary fear and stress. This is not a true test of your abilities and a waste of a valuable resource. Past papers have the greatest value when you have studied the content and are ready to practise the application, testing yourself and finetuning your skills in the near-final stages of exam preparation.

TIP

Be strategic about the order in which you work through past papers.

If you have past papers for 2021, 2022, 2023, 2024 and 2025 – there is some strategy in how you approach these. Our suggestion is:

- Start with a middle one (2022 or 2023) – not the oldest or the most recent.
- Save the most recent one (2025) for closer to the exam (but not TOO close):
  - It will be the most relevant in terms of difficulty, format, and content
  - It will be the most important one to properly test yourself
  - You will be better prepared by this time, giving yourself the best opportunity to do well and enhance confidence
  - BUT make sure you don't leave it too late – if you don't get to it, that will disadvantage you.
- Do the oldest one last (2021):
  - It will be least representative of your actual exam
  - If you run out of time and don't get to it, it won't be a catastrophe

A few points to note:

- Always ensure you have **both** answers and worked solutions for the papers

    - If you can't mark your attempt, or you can't see the full working for each question, it will limit the value of the exercise and create stress.
- Generally, the past papers get progressively harder each year.
- Try **not** to do past papers the day (or night) before an exam:
    - If it highlights a gap in your knowledge, you don't have much time to fix it
    - If you need clarification on something, you don't have much time to get it
    - If you find it very difficult or you can't do something at the last minute this can cause undue stress and knock your confidence before the exam.

TIP

Always try to do past exam papers under strict exam conditions.

Past papers have the greatest value if they are done in conditions that mimic exams.

When doing a past exam paper try as far as possible to:

- Do it in one go
- Do it strictly within the allowed time (use a timer, watch or clock)
- Don't refer to any notes
- Don't look at model solutions or answers
- Handwrite your responses rather than typing
- Do it at the same time of day your real exam will take place
- Do it in a quiet place where you won't be disturbed (if possible)
- Put away all devices and distractions (phone, laptop, etc.)
- Have all necessary equipment with you (calculator, ruler, pens, eraser, etc.).

There may be differing opinions on how to approach past papers, and some people may advocate breaking the paper into parts, treating it as a series of practice questions to be done over time rather than an exam to be completed in one sitting. In our experience this is not ideal, although there may be some exceptional circumstances when you need (or choose) to approach a paper in this way:

- If you do not have a sufficient chunk of time available, then doing the questions ad hoc is still better than not doing them at all. However, with proper planning, this situation can hopefully be avoided.
- If you have a particular weak area that you want to focus on, you may decide to select certain questions to practise, but this would be the exception and not the rule.

In our experience, replicating exam conditions is more beneficial because:

- By doing the practice exam **in one go** you train yourself to focus for the duration of an exam:
    - This could be anything from 45 mins to 3+ hours
    - Exams in early high school tend to be shorter and limited to class time, while in Years 11 and 12 they can be 2–3 hours long
    - If you haven't practised and built up your stamina, it might be hard to sustain focus in a real exam situation.
- By **not referring to notes** you truly test your knowledge and identify gaps to address:
    - If you reference notes to assist in answering questions, you won't get a true gauge on how well you understand and remember the information, and this can create a false sense of security
    - If you haven't truly tested yourself, you may be unaware of gaps in your knowledge which, if only discovered in the exam, will compromise you.
- By **not peeking at solutions**, you train your brain to solve problems without hints.

  - Looking at solutions to help you solve a question, or even to check you are on the right track, deprives your brain of the opportunity to figure it out, which negates some of the learning and can undermine you in an exam.
  - If you look at solutions for guidance you can't be certain whether you would have been able to do it by yourself, and your results may not be an accurate representation of your performance.
  - If you haven't struggled a bit, you may not be maximising the learning opportunity.
  - Checking answers takes time and disrupts flow, diminishing the value of the exercise, especially regarding developing focus and time management skills.
- By doing the practice exam under **strict timed conditions** you learn how to manage your **time, accuracy** and **stress** levels effectively within the time constraints:
  - This benefit applies to the overall exam and individual questions
  - Completing a question or exam under timed conditions is different to completing it at leisure. If you practise exam questions without time constraints, it can yield false or inflated results.
    - For example, in an exam you may only have a few minutes to answer a difficult question, and if you allow yourself an hour to complete this question when practising, you may get the answer correct, but you won't know if this would have been achievable if you only had a few minutes.
    - The same goes for a whole exam – if the exam limit is one hour and you take two, you won't know whether you would have been able to complete the paper or get the answers correct if you only had one hour.
  - To get a true read on your competency and ability to manage time and stress, you must practise under those time-constrained conditions.

- If you stick to timed conditions in a practice and can't complete it, that's a crucial thing to know, and you will be able to work on developing time management skills before sitting the exam.
- It's possible for smart, hardworking students to come unstuck in exams, unable to finish due to poor time management.
- When working on **time management** skills in your practices, here are a few techniques you can apply:
    - Be guided by the mark allocation for questions, distributing your time accordingly. A 4-mark question will usually be more complex and require more time and working than a 2-mark question. So don't get caught up spending a disproportionate amount of time on a 1-mark question, when there are 7-mark questions ahead. This often trips students up.
    - The general guideline for time allocation in exams is 1.5–1.8 minutes per mark. Try to adhere to this in the practice sessions, ideally using less time on multiple-choice and short questions, thereby saving time for longer or more complex questions.
    - The guideline will be different for subjects that have a single question with a large mark allocation, for example, an HSC English essay is 40 minutes for a 20-mark question, but this wouldn't usually call for a 'mark per minute' approach, as the aim would be to complete the essay within the time, having written an appropriate amount.
    - If you struggle with a question, it's best to leave it, circle it so you remember to come back later if you have time, and move onto other questions. Staying stuck for too long means you may run out of time for easier questions later.
    - In a related point – beware of the **'sunk cost fallacy'**, which is the tendency in an exam scenario to persevere

on a question despite having already spent a long time on it, just so the time already invested is not deemed 'wasted'. This can be dangerous in an exam. If you have tried to solve a question twice and still can't get it, flag it for review and move on. This is especially relevant for calculation type questions.

- Note: questions don't always increase in difficulty in chronological order, and question parts are sometimes independent – so you may find in maths, for example, that question 12b is easier and independent of 12a – so if you can't do a question, don't panic, just move on to the next part.
- Watch the time so you are aware of how you are progressing through the paper and how much time you have left – for example, if you have a third of the time left and two-thirds of the paper still to do, you know you will run out of time.
- As you become more practised, set yourself interim time benchmarks so you stay on track to finish.
- In essay or extended response questions, track the time to get a sense of how much you can write in the time limit, so that in the exam you can adjust your speed and writing volume if necessary – for an HSC essay you should aim to write approximately 1,000–1,200 words in 40 or 45 minutes.
- If you are required to do multiple essays, ensure you allocate sufficient time for each – there is no point in having a fantastic first essay and an incomplete second essay. This is a VERY crucial time management skill to master in your practices, because in the HSC English Advanced course, for Paper 2 you will have two hours to complete three essays, and you need to allocate approximately 40 mins to each. In the history HSC exam, you will have three hours to complete a combination of essays, plus short-answer and extended-response

questions. It's ok to spend slightly less time on sections you know you can do faster, leaving more time for sections which might take you longer, but this needs to be rehearsed and intentional. If you haven't had sufficient practice, poor time management can severely compromise your exam performance.

- o Timed conditions introduce an added element of **stress and pressure** which is hard to successfully navigate without practice:
    - ▪ If you take your time with the practice exam, ignoring time constraints, you will find it easier to stay calm and focused …
    - ▪ BUT that's not the reality of actual exams, which tend to be highly time pressured and intense.
    - ▪ If you have not practised in a way that allows you to experience some of that pressure and potential panic, it can be hard to manage when you encounter it in a real situation.
    - ▪ Many students get derailed in exams, NOT because they didn't study or didn't know the content, but because they panicked under pressure and couldn't think clearly.
    - ▪ If the exam will be time pressured, your practice must be time pressured too, so that you can build the necessary skills to focus and function well under pressure.
    - ▪ Rehearsing exam conditions helps you stay calm on the day.
- o Timed conditions help you develop the essential skill of balancing **speed and accuracy.**
    - ▪ It's easier to be accurate when you have all the time in the world to complete a question.
    - ▪ It's much harder to be accurate when you are rushing, and you will hear many very smart students complain about the careless mistakes they made because they had to work fast.

        - In reality, most exams require some rushing, and all exams require accuracy – because you only get the marks if you are accurate – so this is a significant issue.
        - Speed versus accuracy is a trade-off because:
            - Slow and accurate means you may not finish in time
            - Fast and inaccurate means you will lose marks.
            - The only way to get better at both speed and accuracy is to PRACTISE!
- **Handwritten** practice responses will prepare you better.
    - o Exams are usually handwritten so this approximates exam conditions.
    - o Typing is faster than handwriting, so typed practices create a false sense of how much you can achieve within the time constraint, especially for essays and extended responses. This will cause problems later!
    - o So, if you choose to type your practice responses, be sure to CHECK you will be able to handwrite everything within the allowed exam time.
    - o Typing allows easy editing and has autocorrect functions which will not be available to you in an exam.
    - o Handwriting your practices improves hand strength, speed and legibility.
- By doing the practice at the **same time of day** as your actual exam, you train yourself to operate under similar conditions.
    - o If your exam will be after lunch when you usually feel more tired, it's best to practise in these conditions to train yourself to accommodate for this.
    - o If you do all your practices in the morning when you feel fresh, you may find it harder when sitting your afternoon exam.
- Doing the practice paper in a **quiet** place, with **no distractions**, with all your **equipment** at hand is quite self-explanatory – the more closely your practice conditions resemble real exam conditions, the more valuable it will be.

TIP

Always mark your practice exam papers.

The value of doing a past exam paper is limited if you don't mark it. It's only through the process of marking your answers against model solutions and marking guidelines that you truly learn and refine your skills. You can only progress and improve in future if you can identify gaps in your knowledge and diagnose what you got correct (and why) and what you got incorrect (and why). The aim is to fix the things that lost you marks and repeat the things that won you marks.

Caveat: For essay questions and extended responses, especially for English and other humanities subjects, it can be difficult to self-mark. In this case it would be ideal if your teacher could mark your work and provide feedback. Failing that, the next best solution would be peer marking or asking a tutor or family member to assist.

TIP

For calculation questions – look at full worked solutions, not just the answers.

When marking calculations, particularly in maths and some science questions, it can be tempting to skip over the worked solution, marking yourself based solely on the final answer. Although this saves time, it diminishes the learning and can be misleading. As we established previously, you can get a correct answer and not achieve full marks if the progressive working doesn't tick the boxes. If you don't take the time to work through the full solution you may not realise that you would have lost marks, and without this knowledge, you would not know to make the necessary adjustments in the proper exam.

There is another reason why you should look through the full solution. Even if you got the question correct and included all the appropriate working, sometimes there is more than one way to solve a question, or more than one acceptable answer, so it's worth expanding your toolkit, giving yourself more options for answering future questions which may require the use of alternative methods.

TIP

Mark your past papers as soon as possible.

If you mark your practice paper soon after completion, the process is more efficient, saving you time. If you leave too much of a gap before marking, it will take longer, because you have to refresh your memory on what you wrote and why.

TIP

Make a note of question numbers you might want to revisit or redo.

As you mark the paper, it's wise to make a few brief notes for questions you wish to review, query, or retry before the exam. Noting the relevant question numbers and reason for review will save time – it can be hard to retrace your steps later if you need to search through multiple papers to find them.

You can make a note on the cover of each past paper or use a notebook to keep a list. We found it easier to note it on the front of the practice paper rather than reconciling a separate list. Here are some examples of what you might like to make a note of:

- Questions you don't understand
- Answers or worked solutions you don't understand

- Concepts you need to clarify with your teacher
- Questions you got wrong that you may want to attempt again later
- Tricky questions you want to revise (whether you got them right or wrong)
- Questions using unusual or alternative methods

For questions that you got wrong: redo them as many times as necessary until you get them right! It's fine to mark them and see where you went wrong, but you need to be sure you can eventually get them correct without help.

<u>TIP</u> and a CAUTION

The value of practice is limited if it doesn't lead to strategic reflection and correction!

The act of 'practising' is crucial but will only translate into better results if students critically evaluate their performance, recognising the weak spots and creating strategies for future improvement.

This seems like a good time to quote the basketball icon, Michael Jordan, who said, "You can practice shooting eight hours a day, but if your technique is wrong, then all you become is very good at shooting the wrong way".

It's important to bear this in mind because hours of practice repeating the same mistakes can amount to wasted time with limited gains, which is the antithesis of working smart. The optimal approach is therefore to combine a good volume, variety, and quality of practice with detailed marking and strategic reflection. If the learnings are internalised and applied to future work, improvement will follow.

***

While it's very important to apply the principles of reflection and correction to each practice, the next step is evaluation of your performance from a macro perspective. This means taking a step back and looking at the big picture to identify and remedy broader weaknesses or **error patterns** across all your practices.

4. Reflecting on your holistic performance

If each practice task and past paper is viewed in isolation, you risk missing broader weaknesses or error patterns that might need addressing. However, if you take a holistic view across all your practices, you will be able to spot **error patterns** which might be adversely impacting your performance. Once you have recognised these patterns, you can put strategies in place to fix them, which will translate to better results.

As you work through practice questions, revision, and past papers in preparation for assessments, you may pick up patterns of mistakes. For instance, you may notice specific types of questions you get wrong, or perhaps simple careless errors that you repeat frequently. In fact, you may be making these same mistakes in your real exams, across multiple subjects. Both my kids at one time or another made careless mistakes in their practices and exams, which cost them marks. If this happens, don't lose heart. Every student, including top students, will have this experience at some point. The power lies in identifying these patterns in time to fix them.

Before moving on, here are a few examples of common mistakes we have encountered.

- In maths:
    - o Forgetting to correctly apply the order of operations – remember: first do what's inside the brackets, then exponents, then multiplication/division, then addition/subtraction

- o General calculation errors (often due to rushing)
- o Mental maths errors (including basic maths)
- o Accidentally adding instead of multiplying (or using any incorrect operation)
- o Forgetting to write the formula before substituting
- o Transcription errors from the question
- o Transcription errors from one line of working to the next
- o Misreading your own handwriting
- o Accidentally dropping a minus sign from in front of a number or bracket
- o Accidentally dropping brackets
- o Mixing up 'greater than' and 'less than' signs
- o Forgetting to apply basic inequality rules
- o Forgetting to apply basic fraction rules
- o Forgetting to apply basic index rules
- o Forgetting that the square root of a number can be +ve **or** –ve
    - ▪ Except if it's a length or distance, which will always be +ve
    - ▪ Except if specified that it must either be +ve or –ve
- o Making errors by omitting working or trying to do too much in your head
- o Using incorrect unit labels (e.g. cm instead of mm)
- o Incorrectly converting between units of measurement
- o Failing to spot varying units in a question that must be converted into comparable units before calculating – sometimes these are deliberately put into questions as traps (e.g. if a question has both ml and L, both need to be converted to either ml or L before calculating anything further)
- o Failing to present the final answer in the terms demanded in the question – this might be using a specific unit of measurement, rounding to a certain number of places or significant figures, etc.
- o Mixing up perimeter, area, or volume formulae
- o Errors with rounding up or down, or estimating

    - Using the wrong calculator setting
    - Not reading the question properly
    - Forgetting to apply or address ranges or limits stipulated in the question
    - Failing to check the solution satisfies all values and features in the question
    - Not applying logic – e.g. you can't get half a person or a negative distance
- In science subjects:
    - Reading results/data incorrectly off a graph or table
    - Plotting data points inaccurately
    - Forgetting to label graphs including title, axes, units, scale, key
    - Simple calculation errors
    - Using incorrect unit labels (e.g. grams instead of kilograms)
    - Incorrect unit conversions
    - Failing to spot varying units in a question that need to be converted into comparable units before calculating – sometimes these are deliberately put into questions as traps (e.g. if a question has both g and kg, both need to be converted to either g or kg before calculating anything further)
    - Failing to present the final answer in the terms demanded in the question
    - Forgetting to use specific scientific terminology
    - Forgetting to include diagrams
- In multiple-choice questions for any subject:
    - Knowing the correct answer but accidentally selecting the incorrect option
    - Confusing your question numbers – attributing the correct answer option to the wrong question number
- In all subjects:
    - Forgetting to respond appropriately to the directive terms, for example:
        - Not answering the 'extent' part in a 'to what extent' question

    - Not giving similarities **AND** differences in a 'compare' question
    - Not giving reasons in a 'justify' question, etc.
- o Forgetting to answer the second part of a question
- o Forgetting to provide an example when asked
- o Not reading the question properly all the way to the end
- o Overlooking additional instructions
- o Forgetting to check your solution/answer satisfies all aspects of the question

These lists are not exhaustive but are intended to illustrate the types of mistakes students commonly make in both practices and exams. You will notice that very few of these errors would stem from lack of knowledge or conceptual understanding, and if they did, then this should trigger the realisation that knowledge gaps need to be filled. Instead, most of these errors occur because students are under time pressure and **rushing** and are therefore mostly categorised as **careless** errors.

Careless errors are often trivialised but cause immense frustration and upset for students, because despite excellent understanding and thorough preparation, they needlessly lose marks on things they know how to do. These mistakes can unfortunately be costly, and we know a top performing student who lost the maths prize two years in a row by losing a single mark in each final exam due to careless error. This can be soul crushing, so it's worth finding ways to reduce or eliminate these errors on exam day.

The bad news: Everyone makes careless mistakes sometimes (including top students).

The good news: With awareness and practice, most of these can be eliminated.

TIP

Keep a written record of traps and mistakes!

My kids each kept a 'book of traps'. This was a written record of **traps** they fell into and **mistakes** they made. It was not an exhaustive list of every error, or every question they got wrong. It contained their broad reflections on difficult questions and error patterns, particularly 'careless' errors, which they reviewed prior to exams as a reminder of what **to do** or what **not to do**. This served as a learning tool for refining and improving performance. Having a dedicated book with everything recorded in one place made it easier for studying. For example …

In Year 8 maths, both my kids at some point forgot that when you find the square root of a number, the answer could be either positive or negative. E.g. $\sqrt{4} = \pm 2$.

Both made the careless mistake of providing only the positive solution (2) in their answer, because they neglected to show the ± sign. So, this was added to the list of traps which was revised before exams and the mistake was never repeated.

This technique of recording and reviewing common traps and mistakes is extremely effective, keeping them top of mind when practising, redoing, and revising, thereby reducing the potential for similar errors to be made in exams. It's especially helpful for subjects such as maths and the sciences, which have a higher incidence of careless errors, but can be used for other subjects any time you become aware of an error pattern which can be remedied to improve future performance.

***

By this stage, you would have done almost all the hard work and study necessary to confidently sit your exam or complete your assessment task, but there are a few more crucial finishing touches required to complete the preparation ….

5. Finishing touches and final preparation

The importance of final preparation should not be underestimated. Even if you have ticked all the boxes up until this point, success is not assured if you falter at the final hurdle. This last bit is where it all comes together and translates to marks.

Because these final steps often take place just before an exam or task is due, there is a reasonable chance they will fall into Quadrant 1 as they are both important and urgent. If you have planned well in advance, this should already be accommodated in your schedule, but if not, then these tasks should become top priority.

Depending on the nature of the task and how much is already prepared, many of these tasks should (hopefully) not take long, but do not be fooled … they still mostly belong in the most valuable 20% of your effort. Just be mindful of recognising when enough is enough, resisting the temptation to pursue perfection with diminishing returns.

Let's look at some of the final preparation that may be required for various tasks:

- For a speech or verbal presentation:
    - Ensure you have answered the question and addressed any directive terms
    - Finalise your script
    - Ensure your word count allows for a comfortable speaking pace
    - Prepare palm cards (or prompt cards) with highlighted key words:
        - This is your safety net in case you forget what to say
        - Make sure they are well spaced and legible
        - Ideally you should not need to refer to them
        - If you do refer to them, try not to read from them extensively

        - Students will do better if they are less reliant on the prompts
    - Rehearse as many times as possible until you can deliver confidently
    - If possible, rehearse to someone who can provide feedback
    - Record yourself and watch it back to detect areas for improvement
    - Remember to focus on non-verbal communication such as:
        - Body language, hand gestures, and movement
        - Eye contact
        - Tone and volume of your voice
        - Pace of speech
        - Facial expression
        - Often marking rubrics will have specific non-verbal criteria
    - For group work – rehearse together if possible
    - If there is an accompanying slide presentation – ensure it is finalised, edited, proofed and fits with your verbal presentation in terms of timing and content
    - Ensure you can stay within the time limit
    - Ensure you have any required props ready
- For a recorded video presentation or multi-modal task, ensure that:
    - You have answered the question and addressed directive terms
    - The script is finalised and the video is created with voiceover
    - The video and audio align, and audio is clear
    - Any background music does not overpower speech
    - All text appearing in the video is proofed
    - Transitions and effects work
    - Volume and speed are appropriate
    - It adheres to the time limit
    - You are clear on submission instructions
    - Marking criteria are adhered to – they sometimes reward creative use of varied multi-modal tools such as text, images, music, video clips, infographics, transitions, and effects
    - Overcrowding or dense text on slides is discouraged

- For a performance (e.g. music):
    - Rehearse, rehearse, rehearse and get feedback if possible
    - Record yourself and watch it back to identify areas for improvement
- For an oral exam (e.g. languages):
    - Conduct a final review of your notes
    - Ask someone to verbally test you
- For hand-in tasks such as essays, reports, or research projects, ensure that:
    - You have answered the question and addressed directive terms
    - The format and structure are appropriate and correct
    - The correct template has been used (if required)
    - You have proofread and edited to correct spelling and grammar
    - You have adhered to the word limit
    - A bibliography or process log has been included (if required)
    - You are clear on submission instructions
- For a written exam:
    - Conduct your final revision of notes
        - Hopefully, you would already have revised a few times
    - Review your previous exams, tests, or assignments
        - Check you have applied teacher or marker feedback
        - Check you can do questions you previously got wrong
    - Check through your practice past papers and revision
        - Revisit questions you had noted for review
        - Retry select questions you previously got wrong
        - Ensure all queries have been addressed and clarified
    - Review your traps and mistakes lists to refresh your memory
- For **all** types of assessments and exams:
    - Re-read the assessment notification and any marking criteria carefully **before** submitting a task or sitting an exam – ideally the day before
        - This is a crucial step as it's the final chance to catch anything you may have missed, while you still have time to fix it

- o Review the directive terms and their requirements
- o Scan over syllabus dot points or checklists provided to you
    - ▪ This allows you to identify last-minute gaps in preparation
    - ▪ It also creates a final opportunity to test your knowledge on all relevant points
- o Ensure you have all necessary equipment (e.g. spare calculator and pens)
    - o Ideally pack this the night before, because in a last-minute rush you may forget something essential (like a calculator for a maths exam).

***

Most of what we have talked about so far in this section has been broadly applicable to all subjects, but there are some things that relate specifically to certain subjects regarding assessments, preparation, and exam techniques, so in the next section we will provide subject-specific tips and strategies for success.

6. English strategies for assessment preparation and exam techniques

Since *English is compulsory for HSC, and supremely important for the entirety of high school, it makes sense to prioritise specific tips for achieving its mastery.

*I want to reiterate that I am not an English teacher and the aim of this book is not to teach subject content or cover the curriculum. My goal is to share a few key learnings, tools, and strategies that helped us navigate the system to achieve the best outcomes.

Therefore, in this section, I will outline a mix of general study advice and tips gleaned from our experience, as well as specific tried-and-tested strategies for excelling.

TIP

Read or watch your set works! If possible, in the school holidays ahead of time.

The English curriculum from Years 7–12 is largely centred around the study of prescribed texts, and the format is usually one text per term for a specific unit of study. A set work can take many forms, such as a novel, play, poetry, graphic novel, or film. Sometimes set works are referred to as **texts**.

Reading or watching the set work is essential! Covering the text prior to studying the unit creates significant advantage, while conversely, students who are not familiar with the text may be at a distinct disadvantage because:

- If you are already familiar with the plot, characters, and have an idea of the key themes, anything you learn in class will make more sense from the start.
- Teachers often expect or assume that students have viewed the set work in advance (especially if it has been assigned as holiday homework), so they sometimes dive straight into analysis. Students who don't have an overview may find it harder to contextualise and make sense of this analysis.
- It can be challenging to keep up with independent reading during the busy term, so some students may never read the full text, which will be problematic for assessments.
- Often teachers don't analyse the entire text, because realistically there isn't enough class time. They instead focus on key parts – so students who have not previously covered it will develop gaps in their knowledge and understanding, which can create confusion, becoming an impediment later.
- Regardless of what the teacher has skipped over, students are examined on the whole set work, and HSC examiners can test a specific extract from anywhere in the text, so the responsibility

ultimately rests with students to ensure they are comfortable with the text as a whole.

  - o We personally experienced this in my daughter's study of an HSC English Advanced prescribed text. The novel had fifty-nine chapters and almost 500 pages, so it would have been hard to analyse the full text in detail in one term. In this case the teacher focused on the first eight chapters, a few chapters in the middle, and a few chapters at the end, meaning they probably only covered 15–20 chapters at most in class. However, the HSC exam required knowledge of the entire text, and the question was based on one specific extract, which meant if a student had not read the whole text for themselves, they would have struggled to answer the question.

- There is a degree of subjectivity in the determination of what is emphasised in analysis of a text, and there is also variation in skill levels of teachers. Therefore, in our experience, there can be substantial variation between classes, even within the same school, in terms of what is taught and how it's taught. These factors further increase the responsibility of students, particularly in Stage 6, to be self-sufficient and proactive in their learning.

As a last resort – if for some reason you don't read the prescribed text, the next best thing is listening to the audiobook if one is available.

<u>TIP</u>

Study guides for set works can be useful.

Study guides, although not essential, can be helpful as they generally provide summaries with an overview of the plot, characters, key themes, and context. They are **not** recommended as a substitute for class learning or reading/viewing the text independently, but they can be a helpful additional resource to supplement your knowledge and comprehension, especially if:

- The text is long or complicated
- The teacher is unable to teach the whole text
- You want to extend your knowledge and consider additional analytical perspectives, or acquire extra background or contextual information
- You want to do extra practice questions, which are often included in a study guide
- You would like suggestions for quotes and analysis

TIP

The *No Fear Shakespeare* series is gold!

Most schools and most year groups study plays by William Shakespeare. These are notorious for their complexity of plot, themes, and language. The *No Fear Shakespeare* books (published by SparkNotes) saved us on more than one occasion, especially in Years 7–10, because alongside each page of the original text, is a 'plain English' translation to help students understand what is happening. This can be an incredibly helpful study resource, as it effectively eliminates one layer of complexity, allowing students to focus on the deeper analytical components.

In light of the previous tip to read set works independently in the holidays, it would be challenging to try and decode a Shakespeare play without teacher guidance. Reading the *No Fear Shakespeare* English translation makes this more achievable and beneficial. The series is comprehensive, inexpensive, and available at most bookstores.

TIP

Be familiar with the 'module' or 'focus area' within which you study your set work – this applies to all years and especially Year 12.

Most English study units are centred around prescribed texts, with schools having some flexibility in their choice of texts. The way in which you approach and interpret the set work depends heavily on the nature of the unit you are studying, for example, some units focus on analysis of literature and others focus on teaching students how to produce their own creative work. Your success in English will therefore largely depend on how well you understand the conceptual focus of each unit as it applies to the text.

Teachers (hopefully) explain the **conceptual focus** thoroughly at the start of each unit, so that when the set work is taught students understand the framework within which they are operating and what will be required of them. These requirements are integral to the way the unit is assessed and the associated marking criteria.

NESA publishes frameworks or rubrics for each study unit, and it's **advisable for senior students**, and **essential for Year 12** students to proactively seek this information out, ensuring they have a sound grasp of the requirements. This is particularly important with the transition to the new syllabus, which has introduced a few changes.

Note one of the significant English syllabus updates for Years 11–12:

- The current/old syllabus is organised into '**modules**', with each having its own description of what students need to know and do.
- The new syllabus replaces modules with '**focus areas**' and provides a streamlined set of outcomes and expectations that apply to all focus areas.
- These focus areas are individually centred on the same conceptual explorations as the original modules, but the skills being taught and assessed are now consistent throughout all areas.

For example: the new syllabus HSC English Advanced course has four **focus areas:**

- Texts and human experiences (previously *Common module*)
- Textual conversations (previously *Module A*)

- Critical study of literature (previously *Module B*)
- The craft of writing (previously *Module C*)

Each **focus area** has a different purpose and requires students to study and interpret the content from a different angle. For example, in *'texts and human experiences'*, students study their set works to develop a deeper understanding of human experiences, while in *'textual conversations'* students compare texts to develop appreciation of the evolution of ideas and perspectives over time. This is an extreme oversimplification, but hopefully it illustrates the point.

Note: Other HSC English courses have their own focus areas. For further details and updates on your particular English course, I recommend visiting the NESA website.

To assist students, NESA publishes on their website a description of the key concepts and purpose of each module/focus area studied in Years 11 and 12. They also specify the language and terms that are important for understanding and answering questions related to each area, as well as the list of skills students are expected to acquire and demonstrate. Schools use this to inform what is taught and examined internally, and NESA uses this as the foundation for the HSC exams, so familiarity with the specific demands of each module/focus area is essential for success, as this is the lens through which you will study your prescribed texts.

Given the new curriculum is coming into effect, I will keep things simple by referring to **focus areas** instead of **modules** from now on, although the tips and strategies are equally applicable to both the current and new syllabus.

TIP

Ensure you have a BIG PICTURE understanding of your set work.

It's important to note that almost all set works from Year 7 onwards are assessed and examined in an analytical way. In earlier years this may be in the form of an essay, a speech, or a multi-modal presentation, but regardless, it will mostly be centred around analysis. For the HSC trial and final exams, the format is mostly essays, so the following tips will be presented with an essay framework in mind, but most are equally applicable to other analytical formats.

When studying a set work, it's easy to get caught up in the detail, focusing on the micro analysis of quotes, or isolated parts of the text. While the detail is very important, it's even more vital to understand the text holistically, to grasp and appreciate its higher **purpose** and **meaning**. Understanding the bigger picture will be a significant determinant of marks. This is achieved by examining the macro elements within the framework of the **focus area**, and understanding the intentions and **reasons** behind the **choices of the composer**, such as:

- The text or set work **type** (e.g. play, poem, novel, film, etc.)
- The **form** (e.g. dystopian, *Bildungsroman*, tragedy, comedy, ballad, ode, etc.)
- The **characters**
- The **context** or era of the composer (Elizabethan, Victorian, Regency, Romantic, Postmodern, etc.)
- The prevalent **values** at the time of composition (e.g. patriarchal, feminist, etc.)
- The key **purpose** of the composer
- The key **message**
- The **themes** or big **ideas** explored

It's only once students have a thorough understanding of these macro elements that they can attribute meaning in their micro analysis. In other words, analysis that does not take into account the form of the work or the composer's context, purpose, or key message, will be largely superficial, because it does not support or convey the intended deeper meaning.

Therefore, it's imperative in any good analytical essay to consider the **form**, **context** and **values**, composer **purpose** and **key message**, as well as the big **ideas/themes** within the text, and back this up with your **evidence** and micro **analysis**.

Note: In Year 7 you might not be penalised for not explicitly addressing context and purpose, although you may be rewarded if those things were addressed in a meaningful way. However, in the senior years you are **expected** to address these in your analysis, and the absence of these elements will cost you marks.

Given the importance of addressing these macro elements, which can be easily forgotten, I thought it would be useful to provide some strategic tips and prompts. If you get into the habit of including them in the early years, it becomes second nature when it counts later.

TIPS for prompting yourself to include macro aspects in your essays

You should mention the following elements in your **essay introduction**:

- The **composer**, **title** (to be underlined if handwritten), **date**, and **form**, for example: 'In William Shakespeare's tragi-comedy play, *The Merchant of Venice* (1596) …':
    - o The **form** in this instance is **tragi-comedy play**, which is very significant
    - o If you get into the habit of stating the **form** alongside the title, it will become an automatic inclusion in your essays
    - o You will then expand on the analysis of this **form** to shape **meaning** within your body paragraphs.
- The **contextual era** in which the piece was composed:
    - o In this case, the Elizabethan era

  - Then back this up by weaving relevant Elizabethan **contextual values** (such as patriarchal) through your body paragraphs, adding contextual meaning to your analysis throughout the essay.
- The composer's **purpose** and **key message:**
  - For example, in *The Merchant of Venice* these may be along the lines of:
    - **purpose** – to critique oppressive social dynamics
    - **key message** – revealing that marginalisation of minority groups causes social harm.
  - These will drive the development of the **thesis** which binds your arguments
  - Back these key ideas up with **evidence** throughout your body paragraphs
  - By using certain **verbs** you can create **prompts** for yourself which automatically lead to the inclusion of purpose and message, for example:
    - **Critiques, highlights, explores, teaches, warns** are all examples of verbs indicating the composer's **purpose or intent**, and the words that follow these prompts illuminate exactly what that purpose is.
    - **Revealing, conveying, showing, demonstrating** are all examples of words serving as prompts for a **key message**. The words that follow this prompt illuminate the key message.
    - There are many other prompt words that can be used and these are just a few examples.
    - You will notice that the purpose and key message are often intertwined, with the key message being part of the purpose.
- Key **themes/ideas** and **characters** (if appropriate):
  - The conceptual insights, ideas and themes usually link with the purpose and key message and are often implicit in the thesis statement.

- o If the essay follows a character approach, the characters and their representations, choices, or experiences will usually be the vehicle used by the composer to convey their key message and purpose.
- o Characters (if appropriate) are usually dealt with in the final part of the introduction, signposting the main point in each paragraph which follows.
- o These same themes and characters should follow through to the topic sentences of your body paragraphs, guiding analysis which proves the thesis.

> **TIP**
>
> **Be strategic in choosing your quotes and evidence for analysis.**

When analysing set works, we have established that it's advisable to highlight and annotate quotes and techniques within the text itself and create a quote bank which you add to sequentially. By the time you have completed your study of the text, you will have extensive annotations and quotes in your quote bank, so it can be overwhelming and challenging to decide which quotes to focus on when studying or constructing your essays, because you won't realistically be able to use or remember them all.

For hand-in tasks it's a little easier because you can consult your notes to select the best evidence, but it's trickier when preparing for unseen exams, because you will be writing from memory without access to your notes or the text.

With that said, here are some TIPS for discerning which evidence to focus on memorising for unseen exam questions (and some of these can be applied to hand-in tasks too):

- Choose quotes and evidence containing a clear **technique** that can be analysed:

  - o If there is no technique, you will struggle with analysis.
  - o Some common techniques and metalanguage used by students are metaphor, simile, personification, imagery, symbolism, motif, juxtaposition, hyperbole. There are many others which will be taught at school, but these ones occur frequently and are good to look out for.
- Where possible, choose **multi-purpose** quotes which are adaptable and can support various themes, ideas, and arguments so you won't have to remember as many.
- Choose evidence which demonstrates or **supports** the text's **key message**.
- Try not to use quotes that are too literal or obvious – rather find quotes which are more **complex**, allowing you to unpack techniques to add depth and sophistication.
- Choose evidence which is **progressive** and flows logically to support the evolution of your argument.
- Choose a **selection** of evidence from the start, middle, and end of the set work.
  - o You will lose marks if all the evidence is from a small section of the text.
  - o You will always be expected to analyse the 'whole' text, so choose evidence which takes the entire text into account.
- The accepted convention is that evidence within your paragraphs should be presented in the same **chronological** order that it appears in the text.
- If a quote is very long, you only need to remember and use the bits that **prove** your point (provided it contains a technique you can analyse).
- If a quote is long and the bits you need (including the technique) are at the beginning and end of the quote, you need not use the entire quote, instead, you can **splice** the quote or use **ellipses** to include the relevant parts for your analysis.
  - o For example, in *Great Expectations*, one of the full quotes is: "I took the opportunity of being alone in the courtyard to look at *my coarse hands and my common boots*. They had

never troubled me before, but they troubled me now, *as vulgar appendages.*" (2003)

- o The key parts are italicised and can be included in an essay as follows: "...my coarse hands and my common boots ... as vulgar appendages".
- o This captures the key imagery and technique of synecdoche that will be crucial for the analysis, and being fewer words, it's much easier to remember and quicker to incorporate in an essay.

- As a last resort if you forget the exact words of a quote in an exam, you can **paraphrase** or use **key words**. This is not ideal, but it's better than leaving it out if it's instrumental to your argument.
- Not all evidence requires quotes – you can use dramatic **devices** or **form** as evidence for analysis, e.g. comedic tropes or dramatic irony or soliloquy etc.
- In film analysis, you would usually analyse **scenes** rather than quotes.
- You would still have to centre your analysis around the **film techniques** employed by the director/composer.
  - o Some common film techniques are camera angles and shots, colour palette, sound, body language and facial expression, motif, symbolism, visual imagery, lighting, framing, and many more which you will study in class.
- When analysing a film, you usually **combine two or more techniques per scene** to convey the meaning, for example:
  - o Lighting and sound
  - o Visual motif and camera angle.

TIP

Be strategic in your preparation for unseen English essay exams.

Studying and preparing for English essay exams can be extremely daunting, because texts can be long and complex with multiple characters, themes, ideas, and additional layered elements such as purpose, context, values, form, and the expectations of marking criteria or the NESA focus area. There are numerous interpretations and considerations, and the exam questions are mostly unseen, making preparation an overwhelmingly stressful experience for many students. It can be hard to prepare for the unknown, but it's worth developing a strategic approach because you are guaranteed to encounter this scenario many times through high school. In Year 12 alone, you will write multiple unseen essays in term assessments, trial exams, and final HSC exams.

Without an effective, targeted approach, you can get lost in the swamp of information, wasting valuable time working hard without producing the desired results. Success in English thus requires both a hardworking and smart working approach, with effort focused on the most valuable 20% of work that will generate most of your results. Your English work will be most effective when you operate from Quadrant 2, with consistent skill building and preparation, because urgent Quadrant 1 cramming will be less likely to yield great results.

So … **how** *do* you prepare for an unseen question which could literally ask anything?

As impossible as this sounds, **there is a solution**! Unfortunately, it's a solution that isn't often taught at school, and which we discovered incidentally. This is one of those things I wish we had known in earlier years, because it would have saved immeasurable time and stress, and would also have allowed greater opportunity to practise and build skills. To date, **it is one of the most valuable strategies we have learnt**. The solution is…

TIP

Craft a MASTER ESSAY centred around a MASTER THESIS…

…which you will memorise and adapt.

In the same way a master key opens multiple locks, a **master essay** which is cleverly and thoughtfully constructed will be adaptable to multiple questions that may arise. This forms the foundation of your preparation for unseen analytical essay questions. It will take you time to set this up properly initially, but it is worth the effort because, if it's done well, it will generate **significant returns in your results and save you a lot of time** later. This is the epitome of **working smart**, and I believe it's non-negotiable if you want to excel in English.

The idea is to create one **master essay** per focus area (or unit of study), **memorise** it, and **adapt** it for various questions that may arise in an exam. This is especially helpful in Year 12 where you will be examined on all the focus areas in the final exam.

This idea of a master essay is a bit controversial. There are some teachers who might discourage it, but we have found that many teachers, most tutors, and students attuned to working smart, advocate for this method. In our experience, this is a valuable strategic approach which pays great dividends if it's executed correctly. But before we debate the merits of a master essay, let's examine exactly **how** you would create one.

We have found that the best strategic approach is to become familiar with the key aspects of the **focus area** and/or **marking rubrics**, the **purpose** of the composer, and the main **message** and **concepts** in the text. From this you craft a **'master' thesis** and a **'master' essay** which address these aspects in an insightful way. Finally, you back this up with evidence and analysis. This would be beneficial in all years but is especially crucial in Year 12.

This sounds complicated and when you first set out to create a master essay, it can be difficult, but once you have done it a few times, the process gets easier and more intuitive. If you can start applying this method as soon as you are required to answer unseen essay questions (which is likely to be from Year 7), you will be well rehearsed by Year 12 and in a strong position to excel in your HSC English course.

When crafting a master essay that will be used in an assessment, these are a few of the elements to consider and incorporate:

- The main **conceptual focus** and **key terms** contained in the **focus area**, and/or any **marking criteria** issued ahead of time:
  - Unseen essay questions will almost always examine aspects of the focus area and its key terms (especially with the defined frameworks in Years 11–12), so including these in your essay in a meaningful way will automatically tick boxes on the marking criteria – I say, 'in a meaningful way', because throwing in terms without supporting evidence and analysis will not score marks.
  - If you craft a master essay which considers these ideas, you know you are on track to answer a range of questions that may arise in the exams.
  - In earlier years, you will mainly be guided by the key terms on the assessment notification, indicating **what you are being assessed on**, because this will hint at the conceptual focus of the unit you are studying.
- The **form** of the relevant text(s), which will guide your approach.
- The **context** in which the text(s) were created and the associated **values**
  - Helps to inform interpretation of text.
- The key **purpose** and **message** of the text(s)
  - Forms the basis of your insights and thesis.
- The key **themes** or **ideas** conveyed in the text(s).
- The key **character(s)** you will analyse.
- The **structure** of your essay, which often depends on the unit you are studying:
  - You might analyse by **character**, where each paragraph is centred on a character, and the thematic ideas are conveyed throughout.

   - You might have a **thematic** essay where each paragraph deals with a different aspect of a theme – and you might use various characters throughout your paragraphs to convey the ideas.
   - You might have an essay which tracks **one key character** through their arc, with each paragraph signposting a phase of development.
   - You might have a **comparative** essay where you are comparing two or more texts, and you could structure your paragraphs by text or by theme.
   - You might have an essay centred specifically around **macro elements**, such as **form**, **purpose**, or **key message**, where characters are not the focus.
   - The structure will be guided to a great extent by the nature of the focus area and the type and form of texts you are studying.
- Once you have clarity on all these elements you can construct:
   - A broad overarching insightful **thesis**
   - **Topic sentences** containing your main line of argument for each paragraph.
- Then, finally, you build your paragraphs with the **evidence and analysis** supporting your thesis and topic sentences, ensuring that they work together holistically.
- Your essay should adhere to the structural conventions outlined previously.
- Make sure your master essay stays within a reasonable word count so that it can be handwritten within the time limit in an exam.

TIP, cautionary note and an alternative approach

For some units of study you might need to prepare extra **master paragraphs**. For example, if you are studying seven poems in a particular unit, you might only analyse three of them in your master essay, devoting a paragraph to each. However, you don't know whether you would be able to use those same three poems in an exam. Those poems may not be suitable for answering the question, or in the worst-case scenario, an exam question could instruct you to reference a **specific** poem. If you have not prepared analysis for the additional poems, this could cost you dearly. In this case it would be advisable to prepare your master essay with your three preferred poems and create additional standalone paragraphs for the remaining four poems, which could be substituted if necessary. Just ensure they all fit with your master thesis. The same approach could be applied if you were composing your master essay with a character focus. You may want to prepare one or two standalone paragraphs on extra characters, catering for certain questions that may arise. Having said this, be sensible in your approach – there is no need to go overboard creating extra master paragraphs if they won't be required.

Instead of a **master essay**, you may opt to rather develop a series of **master paragraphs** which could be assembled in any order depending on the question. The suitability of this method would depend on the adaptability of your master thesis and might be better suited to some focus areas than others.

***

You may be sceptical about how a single master essay could possibly answer a broad range of questions. To be honest, I was also wary at first and it seemed like a risky approach. But having seen this used extremely successfully by many top-achieving students, and seeing it work across all focus areas, I believe it's a brilliant strategy! The key to the strategy is incorporating the crucial **macro components** such as **concepts and terms from the focus area**, **context**, **values**, **purpose**, **form**, **conceptual ideas,** and **key message** … because these are required for answering most questions and would therefore be at the heart of **any** argument you make. If this is backed up by appropriate **micro analysis**, it follows that the analysis would also be relevant for most questions. The adaptation would then involve finessing your master thesis, topic sentences, and pre-prepared analysis to answer the specific question you are confronted with. If you can do this well, it will save time and enhance quality. The alternative, in the absence of a master essay, would be to study and memorise all notes and entire quote banks, preparing large volumes of analysis on a variety of themes to equip you for creating an essay on the spot under timed conditions, which can be risky and challenging to pull off. This approach would be much more time consuming and less likely to hit the mark.

TIP

DO NOT create your master with a specific question in mind.

The whole point of a master essay is that it is **not** tethered to any specific question but rather created to be **flexible for adaptation**. If you create the master thesis and essay in response to a specific question, you pigeon-hole yourself, essentially 'locking' the essay, because it will contain language and elements specific to that scenario, making it difficult to adapt to a different question later.

It can seem tricky to come up with a thesis for your master when you aren't answering something specific, but if you have considered the central requirements implicit in the focus area, and you understand the composer's purpose, key message, themes, and characters, you can develop an insightful and flexible thesis which incorporates these aspects and can be adapted or applied to fit whatever question arises.

TIP

DO NOT make your master so generic that it lacks insight.

In avoiding a master that's too specific, you run the opposite risk of creating one that's too generic, lacking insight or depth, rather becoming a superficial recount of the text. It can be challenging to create a master which is adaptable while also insightful. The trick is to revisit the demands of the focus area or marking criteria to understand the conceptual essence of the unit and ensure you have a comprehensive grasp of the aforementioned macro elements, which provide what you need to craft a broad insight that is flexible while still containing conceptual substance.

> **TIP**
>
> **Your master does not need to take a unique or controversial stance!**

Students may be under the impression that to stand out they must create an argument that is unusual, niche, or contrarian. While there are a few highly skilled students who might be able to pull this off, it's a risky strategy and can be costly if not executed extremely well. Given that marks are generally awarded for clear, cohesive, and well-substantiated arguments, there is no need to create excess complexity which can backfire. A clear, sophisticated argument aligned with conventional interpretation will usually serve you better than a highly original or niche argument, which may be difficult to sustain or harder to adapt.

> **TIP**
>
> **Leave deliberate gaps in your master essay allowing for adaptation to specific questions.**

When crafting your master essay, be mindful that you will need to **adapt** it when faced with a specific exam question. Think of the master as being 80% of a finished product. It's your foundational essay – not your final essay. It's therefore advisable to keep the master word count around **200–300 words less** than you would need in a final essay, leaving space to adapt and add wording which directly answers the given question. The most strategic places to leave gaps would be in your introduction, conclusion, and topic sentences. I also suggest that at the end of your master introduction, conclusion, and every paragraph, you write the words: 'link to question' to remind yourself that this part needs to be customised to answer the question.

TIP

Ask for feedback and accept that your master essay will go through multiple iterations.

Your master essay probably won't be created quickly or finalised on the first draft – it will take time to prepare and refine. It's advisable to seek feedback during this process, so you may need to redraft multiple times. This can be time consuming and frustrating, but it's worthwhile to create a good quality essay. If the master is well constructed and insightful, it will be a valuable asset, saving time and scoring you marks later. This process contributes significantly to the most valuable 20% of your effort.

A small caveat for students with perfectionistic tendencies: try to avoid spending endless hours making tiny changes in pursuit of the perfect master essay. Once you start chasing diminishing returns, it's important to recognise this and stop, especially because the master is not intended to be the final product.

TIP

In Year 12 you can craft master essays well in advance and independent of assessments.

In Year 12 English, unlike other years, **all** your focus areas and associated texts are examined in the final HSC exam. You will therefore be required to write four extended responses, three of which are analytical essays. I would strongly advise you to begin working on your master essay for each focus area as soon as it's feasible to do so, even if there is no immediate assessment. Once you are familiar with the demands of the focus area and have studied the set work in class, you should have all the information you need and be in a strong position to create a master essay, especially if you have been annotating your text and updating your quote bank progressively.

Proactively creating master essays will provide significant academic **advantage** as:

- You are working in Quadrant 2, which is non-urgent, high-quality preparation.
- The master essay can take quite a long time because as we have established, it should go through a process of feedback and refinement. Working ahead ensures you have given yourself enough time to create a high-quality, refined essay.
- The time and thought dedicated to creating and refining a master essay generate deeper understanding of the focus area and the text.
- When assessments arise, much of the work in the most valuable 20% will already be done, freeing you up to focus on higher level preparation and practice.
- In Year 12 you may be examined on each focus area up to 2–3 times, so preparing a great master essay early that can be used in multiple assessments and exams saves time and creates enduring advantage, with a huge return on upfront effort.

Conversely, **delayed** creation of master essays can result in academic **disadvantage** as:

- If the master is not done ahead of time and lands up in a bottleneck with other work, or even worse, becomes a last-minute rush in Quadrant 1 before an exam, the process is compromised, impacting the quality of work, potentially undermining your exam performance, and creating additional stress.
- If you don't create a master as soon as you finish a focus area, you will move onto other work and struggle to find time to come

back to it. When you finally revisit it, the unit may no longer be fresh in your mind. This scenario increases the challenge and diminishes quality.
- Delaying the preparation of the master means that in the lead-up to assessments you may still be creating initial drafts, leaving minimal time for practice.
- In Year 12, some schools may not examine all focus areas before trial exams, or even in the trial exams, so if you only create master essays as required for specific assessments, you may approach trials or finals with no master essay for some areas. This creates challenges from a time, quality, and stress perspective.

TIP

Memorise AND practise adapting your master essays.

Once you have a master essay which is insightful, refined in expression, and flexible, the next step is to memorise it.

There are a few methods you can use to effectively memorise your essay:

- Read it multiple times
- Speak your essay out loud
- Record yourself reciting the essay and listen to the recording
- Try to write your essay from memory, highlighting the bits you forget
- The approach you choose depends on personal preference and time constraints, but the key to memorising is repetition.

When memorising your master essay, there is also a strategic hierarchy.

- Memorise the **quotes**, **techniques,** and associated **analysis** within each paragraph **fully**, because these will require the least adaptation, as long as you can effectively communicate how they prove your thesis and answer the specific question.

- Memorise the **introduction**, **conclusion**, and **topic sentences selectively** – although you must know them well enough to retain sophistication, these are the bits that will change the most when you adapt, as you finesse them to include **key words** from the **specific question** to fully engage with it in your answer.

However, it's important to remember that memorising is not the end goal …

✗ **The point is not** to simply memorise the essay with the intention of regurgitating it word for word in an exam. This will NOT WORK.

✓ **The point is to** memorise it with a view to **adapting** to any given question.

You may be wondering why you would go to the trouble of memorising something that needs to be adapted anyway? Well, the benefit of memorising a thoughtfully composed and refined master essay, is that when confronted with an unseen question, you can be assured that many of the crucial considerations and terms have already been incorporated. This minimises the risk of forgetting important elements under time pressure and facilitates the retention of clear and sophisticated pre-planned expression. This frees up time and cognitive space allowing you to concentrate on adapting to the specific demands of the question.

Without a master essay, you will need to take time in an exam to mentally sift through and pick quotes from the original text or a large, memorised quote bank, identify techniques, create analysis, form an insightful argument which answers the question, and simultaneously make it sound good. That's a lot to do in a short time! However, if you have already memorised the conceptual thesis, relevant context, quotes, techniques, analysis, and sophisticated phrasing, then you can devote more time and brainpower to answering the question and adapting your essay in the most insightful way. It's best to think of memorising as your rehearsal, and even if you must make significant adaptations in the exam, the rehearsal and preparation will always be more beneficial than taking a gamble and improvising the whole thing on the spot.

Some teachers discourage memorising master essays, because if they are simply regurgitated without appropriate adaptation, this results in a failure to answer the question, which is certainly problematic since answering the question is required to win marks. This is why memorising is only part of the strategy and to be effective, you also need to employ the technique of **adaptation**. The ability to **adapt** a master essay takes **skill**, and as with any other skill, this requires a lot of **PRACTICE**.

> **TIP**
>
> **PRACTISE adapting!**
>
> **DO NOT let your exam be the first time you adapt your master essay!**

Using an untested master essay in an exam is extremely risky. It's critical that prior to an exam containing an unseen essay question, you do a few adaptive practices … the more the better. There are a few different methods to practise this technique:

Write full practice essays

The first method is to find practice questions or past papers for the unit/focus area and set work you are studying and write as many full practice essays as you can, each time using your master essay and adapting it to answer the specific question.

Adapting to a question involves the following key steps:

- Identify directive terms and unpack the question to determine what is being asked.

- Develop your specific thesis to answer the question, utilising as many elements from your master thesis as possible. If it is strong and flexible, you should be able to integrate the key elements of the question into your master thesis, creating an overall thesis which is tailored to the demands of the question.
- Follow a similar process to adapt your topic sentences for each paragraph, incorporating key words and concepts from the question so they support your tailored thesis and ATQ.
- If your quotes, techniques, and analysis within paragraphs are well selected and constructed you should be able to use them straight from your master, as long as you **adapt** and **reframe** the analysis to support your answer to the specific question, which may involve some rephrasing.
- If any of your evidence does not fit the specific question, you may need to substitute for more appropriate evidence, so it might be good to memorise a handful of extra quotes with analysis just in case. (Note: This is a worst-case scenario safeguard – you don't want to switch evidence unless it's absolutely necessary – if possible always try to adapt what you have in your master.)
- Ensure that the end of your introduction, each paragraph, and your conclusion link back to your specific question.
- If there's a stimulus, ensure you incorporate this into your adapted essay throughout.

This is the best and most comprehensive type of practice because:

- It's a true test of whether the master essay works and is adaptable
- Writing a whole essay helps develop skills in adapting all parts of the essay
- Writing out the whole essay assists with the process of memorisation
- It's an opportunity to assess whether the word count is realistic or ambitious

However, this method is time consuming and won't always be possible or practical, which is where the second method can be very effective …

Prepare skeleton adaptations

Applying the **80/20 principle** to the practice of **adapting** master essays – the greatest value undoubtedly lies in the adaptation of the **thesis** and **topic sentences**. This is the most challenging part, and it's also the most crucial for success. There is less value in rewriting the bulk of the essay which requires less adaptation.

To implement this approach, you still need to find appropriate past papers or questions, but instead of writing a whole essay, you focus on preparing your customised thesis and topic sentences as outlined above, and merely check your evidence is still applicable. If this is done properly, you should be able to follow this through in a full essay. The advantage of this method is time efficiency – you can complete a greater volume of practice questions than if you were writing full essays.

Once you have practised this thoroughly and developed your adaptation skills, you may be able to complete this method mentally as a brainstorm without writing it all down. Doing this for increasingly complex and difficult questions is invaluable in building skills for exams that require you to quickly adapt mentally before you start writing.

The best method?

There are clear benefits to both writing full adapted essays and preparing skeleton adaptations. The **best** method ultimately depends on your skill level and time constraints. The optimal approach in our experience is a **hybrid** of both methods. In the lead-up to an exam, it's beneficial to write a few fully adapted essays and supplement this with a variety of skeleton adaptations. This facilitates skill building, while also making optimal use of time to cover a broader range of practice questions.

The key thing to remember, regardless of your chosen approach, is that the adaptation is the secret to a successful essay. If you simply rewrite your basic master essay, it's unlikely to answer the question, and you may not achieve the results you want.

> **TIP**
>
> **In the absence of a master essay – doing practice questions is even more vital!**

We've talked a lot about the benefits of a master essay and the process of creating, memorising, and adapting, but for whatever reason, you may find yourself in a position where you don't have a master essay. In this case, you still need to be familiar with the focus area and macro elements of the text, plus have an extensive quote bank which includes techniques, themes, and associated analysis … and it's imperative that you still practise answering questions. Once again, the ideal situation is sourcing past papers or practice questions and writing full essays, making sure you can adapt your knowledge and ideas to answer the question. This practice of writing full essays will be absolutely critical in the absence of a master, because if you have not practised, the first essay you write will be in the assessment, and that's incredibly risky.

Once you have confirmed that you can effectively compose an essay on the spot, you can progress to the second method of preparing skeleton practices. If you are not working from a master essay the process will include an extra step, because you additionally need to consider what evidence to include. You would therefore create a skeleton essay scaffold including the following:

- Thesis to ATQ
- Topic sentence for Paragraph 1 which ATQ
  - o evidence 1
  - o evidence 2
  - o evidence 3
  - o evidence 4

- Topic sentence for Paragraph 2 which ATQ
    - evidence 1
    - evidence 2
    - evidence 3
    - evidence 4
- Topic sentence for Paragraph 3 which ATQ
    - evidence 1
    - evidence 2
    - evidence 3
    - evidence 4

> **TIP**
>
> **You can create a master for creative pieces too … somewhat …**

In Year 12 English, one of the **focus areas** requires students to craft their own writing, which may be in the form of imaginative, discursive, persuasive, or reflective. In the new syllabus there are no prescribed texts for this unit. Students are taught the process of writing and the key conventions for different genres as they learn to create their own.

The **craft of writing** unit is notoriously tricky to navigate, and students tend to struggle because while they have creative freedom, it lacks structure and is completely open ended. Students are fearful because in the HSC they don't know ahead of time whether they will be asked to write a discursive, imaginative, persuasive, reflective, or combination piece. This makes it difficult to prepare and necessitates the creation of more than one **master** to cater for different scenarios.

For example, in the HSC exam, the question could ask for:

- An imaginative (20 marks)
- A shorter imaginative (12 marks) plus a reflection (8 marks)
- A discursive (20 marks)

- A shorter discursive (12 marks) plus a reflection (8 marks)
- A persuasive (20 marks)
- An imaginative (10 marks) plus a discursive (10 marks)
- A question giving students a choice of form (20 marks)
- Any other combination of genres and mark splits

The 'question' could be a visual prompt, a stimulus, or a direct question. It could be general or thematically specific. It could literally be anything. So how should you approach your preparation?

The best advice we received was to prepare **masters** for the following in advance:

- At least one **imaginative** piece – leaving space for adaptation while also containing flexible bits which could be cut down or modified if needed
- A **reflective** piece to accompany the imaginative
- At least one **discursive** piece – leaving space for adaptation while also containing flexible bits which could be cut down or modified if needed
- A **reflective** piece to accompany the discursive
- A **persuasive** piece – which can be an adaptation of your discursive piece
- A brainstorm and skeleton of at least five more ideas for alternative imaginative and discursive pieces.

As a guideline for Year 12 exams, the following word counts would be appropriate:

- A 20-mark piece would be approximately 1,000–1,200 words
- A 15-mark piece would be approximately 800–900 words
- A 12-mark piece would be approximately 600–700 words

The trick to preparing a creative master is ensuring you leave enough space in your word count to adapt to the specific question and simultaneously create a structure which allows you to sensibly cut some of it out if the question ends up being a 12-marker which requires only 600 words.

The creation of these master documents would follow the same process previously outlined for analytical master essays. They would be drafted, then after feedback, finetuned, memorised, and adapted to the question (if possible). They could similarly be practised using past questions.

The scariest part for many students is the pressure of coming up with creative ideas, as this may not come naturally. They worry that if they can't manufacture an innovative idea or groundbreaking originality, they won't be able to succeed. The good news is that students can still do exceptionally well even if they struggle with the creativity part.

We learnt the very valuable and reassuring lesson that success in this focus area depends less on creativity and much more on applying the formula of convention.

TIP

When crafting your pieces, focus on conventions first!

We learnt that the topic, plot, or storyline is less significant than the inclusion of appropriate conventions for each creative writing form. This means students with highly original and creative ideas could achieve disappointing results if they didn't include the conventions the markers were looking for. It also means that less original storylines could score top marks if students included the most important features of their form. In this unit students are assessed on the application of creative writing skills learnt in class. This mostly involves the purposeful inclusion of various structural, stylistic, and language techniques – this is where the creativity lies, and it's good news for students! The key to success is understanding the criteria that need to be met, and once you know what they are, the task becomes less overwhelming.

Here is a brief summary of the main requirements for each form of creative writing:

- Imaginative writing TIPS:
  - Create an engaging title
  - Create an engaging opening to hook the reader
  - Clearly convey **your** purpose or intent
  - Explore a theme or idea, rather than making it plot driven
  - Create a clear setting for the reader to connect with
  - Create a clear character for the reader to connect with
  - Include a range of language and literary devices (metaphor, imagery, symbolism, motif, etc.)
  - Create a deliberate structure for example:
    - Linear →
    - Present – flashback – present (this structure is one of the easiest to trim down if the question commands a shorter response)
    - Introduction → complication → resolution
  - Structure with paragraphs or distinct sections of text
  - Use varied sentence structure and syntax
  - Be coherent
- Discursive writing (which is like a discussion) TIPS:
  - Create an engaging title
  - Explore a theme or idea
  - Show multiple perspectives
  - Use a conversational tone
  - Pose questions (usually rhetorical)
  - Include personal anecdotes to introduce ideas (can be used as an initial hook)
  - Include figurative language and imagery (e.g. metaphor, motif, etc.)
  - Include references to broader reading (intertextual reference) if you can, but not related to other set works

  - o Structure paragraphs (including use of subheadings)
  - o Create a cohesive flow
  - o End with a question or reflective statement, not an absolute
- Persuasive writing requires you to:
  - o Create a persuasive title
  - o Present a clear thesis upfront and keep returning to this throughout
  - o Advocate strongly for one point of view
  - o Use rhetorical questions
  - o Use highly modal and persuasive language
  - o Present statistics, anecdotes, quotes, and evidence to convince the reader
  - o Use a clear paragraph structure, each with a main claim
  - o Use emotive imagery
  - o Summarise and restate your position
- Reflective writing requires you to:
  - o Support and justify your choices in the imaginative or discursive piece
  - o **Analyse** your own writing
  - o Quote yourself extensively and analyse how your choices and techniques convey your intended meaning and purpose
  - o Use terms such as: 'I chose', 'I decided', 'I conveyed', etc.
  - o Include evaluative language such as skilful, effective, etc.

The application of these conventions will demonstrate to the marker that you have developed a sophisticated comprehension of the **craft of writing**.

In addition to the application of conventions, here are a few other tips we have learnt which will help to elevate your writing in this focus area ...

TIP

Use aesthetics to add creative impact.

Don't be afraid to play with aesthetics – artistry in the visual presentation and structure of your writing can be extremely effective.

Some examples (depending on the genre of your writing) might be:

- Varying text alignment and indents
- Non-uniform spacing to make things stand out
- Standalone lines for emphasis
- Using only the left or right side of the page to differentiate between narration and a character's thoughts
- Inserting dialogue
- Using symbols (e.g. to demarcate jumps in time)
- Putting in timestamps to illustrate the passing of time
- Writing selectively in capital letters
- Using subheadings

TIP

Show don't tell … and use imagery.

I am sure that most students are familiar with this concept, because it is the mantra of most English teachers. For those who may not have heard it before, the idea is that instead of saying something outright, it's more effective to create a vision or feeling to convey your message to the reader.

For example, instead of **telling** the reader directly that a character is old, it's better to **show** this in a vivid description of wrinkled skin, thinning hair, or stooped posture.

It would be even better to use imagery and figurative language (such as metaphor or simile) to convey ideas in a more sophisticated and abstract manner.

TIP
Evoke emotion in the reader.

The best way to add depth to your writing and connect with the reader is to make them **feel** something. This can be achieved through your ideas, characters, setting, and your use of techniques such as emotive language, imagery, metaphor, symbolism, etc.

TIP
Take care not to overwrite.

In implementing **show don't tell** and **evoking emotion**, be careful you don't swing too far in the other direction and overwrite. You don't need to go overboard or cram too many devices into a simple description. Sometimes less is more, and sophistication and substance can be found in simplicity.

TIP
Focus on a moment in time for imaginative pieces.

In our experience, vignette-style imaginative writing that goes into greater depth on a smaller fragment of time works better than a plot-driven piece which superficially spans a broader timeframe. Smaller moments in time help you avoid the temptation to outline a sequence of events, prompting you to delve deeper into character, setting, and emotion, generating substance and lending a richer quality to your writing.

If your piece requires an evolution over time, using tools such as 'jumping forward' or 'flashbacks' allow you to deal with the passage of time, while retaining depth.

TIP

Avoid overused or clichéd topics.

In the HSC, there are certain topics and settings which have been overused and become clichéd. We were advised to avoid settings such as schools, bookshops, or libraries, and to avoid ubiquitous topics such as 'school students doing the HSC', 'exam stress', 'when I was in kindy' and 'artificial intelligence' to name a few. Markers are hoping for students to explore less familiar settings and topics, showing evidence of **ideas** and **understanding** beyond the expected.

TIP

Tap into timeless, universal themes and ideas.

If you have a strong central **idea**, it enhances the overall quality of your writing. Additionally, tapping into timeless human and existential concerns such as belonging, connection, fragmentation, memory, truth, identity, freedom, passing of time, mortality, etc. makes your writing more relatable for the marker, and clearly establishes your authorial **purpose**. Incorporating universal themes also gives you significant flexibility, making it easier to adapt in response to a specific question or prompt.

TIP

Make smart use of a central motif.

A **motif** is an element which recurs throughout your writing and contains symbolic significance. Incorporating a central motif elevates your writing, helps thematic development, and ticks some of the convention boxes. Some good examples would be light and dark, or cold and warm. A motif could also be an object, image, sound, or action to which you attach significance. Motifs can be used effectively in both imaginative and discursive styles.

Another benefit of a clever motif is that it helps make your writing more adaptable to a variety of questions or prompts, which brings me to the next tip …

> **TIP**
>
> **Remember, above all you still need to adapt to answer the question!**

The creative writing component of the exam will also have a question or prompt that needs to be addressed or answered. As we have already established, this can be one of the most difficult sections to adapt, because your writing can be on anything, the question can be on anything, and the chance of these aligning exactly is slim.

To give yourself the best shot at an **adaptable creative master**, try to build it around:

- A **central idea** that is **relatable** and **flexible** (as per the examples of timeless and universal themes in a previous tip)
    - For example, the idea of connection/disconnection could be intelligently applied in relation to family, society, culture, technology, mental health, and various other interpretations which you could mould to fit many questions.
- A **flexible central motif or metaphor** which you can work into various scenarios.
- A **flexible personal anecdote** that can be adapted for creative or discursive use or used to convey a range of ideas.

Just as with the analytical questions, your ability to **answer the creative question** will ultimately determine your success. The best piece of writing will not help you win marks if you don't answer the question. The practice of adapting your piece using past questions for this focus area is therefore even more crucial, because it will test your writing against a variety of possible scenarios, giving you a true read on its flexibility.

If you can cleverly twist your pre-prepared master in a way that answers the question, this is ideal, but in the unfortunate case that you are unable to adapt your work, you will need to write on the spot. Having brainstormed alternative ideas will be a tremendous help if you need to write a new creative piece from scratch in an exam, because you will already have generated a potential framework. Although this can feel frightening, all the same rules apply, and if you remember to include the conventions and employ the tips outlined above, you will be able to create a strong piece.

***

To wrap up this section, I want to reiterate and summarise the value of preparing master essays and creatives, especially in Year 12 …

Although the initial preparation of 'masters' can take time, it generates significant benefits in the lead-up to exams:

- The upfront work facilitates deeper familiarity and understanding of the unit …
- … and saves valuable time, reducing workload in the critical period before exams.
- A master is prepared once and can be used multiple times throughout the year – meaning the upfront investment pays ongoing dividends.
- Each time a master is used in an exam, teacher feedback can be directly applied to improve for the next exam.
- Memorising a few essays and paragraphs is much quicker than going back to source documents and memorising pages of quotes, notes, and background information …

- … and analysis prepared *in situ* in a master essay is easier to remember and adapt than that which exists in isolated fragments in quote banks.
- A master with an insightful thesis prompts conceptual and structural cohesion, enhancing analytical flow and clarity of ideas, which is transferable to exams.
- The practice of adapting the same master to various questions enhances skill and memory and is much easier (and more helpful) than writing multiple one-off pieces.
- If you can easily adapt your master in an exam, you will be glad you did the work…
- … and in the unlikely event that you can't adapt and need to write on the spot, you will still be able to use at least some of your pre-prepared piece, retaining elements of rehearsed sophistication.
- A good quality master automatically includes all the key elements, freeing up cognitive space in exams to focus on ATQ …
- … but in the absence of a master, there are infinitely more **on the spot** considerations, making it more difficult to execute well, increasing the likelihood that a crucial element may be forgotten, and potentially reducing the focus on ATQ.
- Familiarity with a master facilitates speed, saving time in exams.
- Having the assurance of a master reduces exam stress and increases confidence.

For students who want to **work smart**, **save time**, and **succeed in English**, creating **master** pieces is strategically essential. This engages with the principles of good time management, quality preparation in Quadrant 2, and the 80/20 rule. If you remember to adapt to answer the question, there is no downside.

An important caveat …

Although I have provided an enormous amount of information and direction around the creation of **masters**, I feel it's necessary to moderate expectations. It would be unrealistic to expect that reading through this once would enable you to sit down and effortlessly write a beautiful Band 6, top mark essay in an hour. In reality, the creation of master essays takes time, a bit of struggle, and a lot of refinement, but my hope is that these step-by-step guidelines and tips, which detail **what** to do and **how** to do it, will speed up and smooth the process. At the very least, this serves as a strategic checklist for you to reference along the way, ensuring that all necessary considerations have been factored in.

Note: Although this section on the preparation of master essays has been focused on English, the same principles can be applied to many of the humanities subjects. In most subjects which entail extended responses or essays, it's possible and recommended to pre-prepare **master essays** or **scaffolds** which can be memorised and adapted in exams.

TIP

For Year 12 – keep your master essays fresh in your mind – especially after trials.

Many students prepare a master, memorise it for an exam, and then promptly forget about it. This is fine for most years, except Year 12. In Year 12 you are examined on the content a few times throughout the year, including in trial exams and HSC final exams. It's therefore worthwhile to memorise your completed master essays throughout the year, keeping them fresh, otherwise you will find yourself trying to remember multiple pieces at the last minute, which makes memorisation more difficult and less effective.

Even just reading through your essays a couple of times a week on a rotational basis will generate cumulative gains. It does not take long but creates significant advantage and makes your study much easier in the approach to final exams.

***

Before moving on, here are some final tips for mastering English and humanities subjects …

TIP

Always plan before you start writing – especially in an exam.

Essay exams are time pressured, and because students are concerned about running out of time, they tend to read the question and launch into writing the essay immediately. This can backfire, because if you have not taken time to plan how you will answer the question, and create or adapt your thesis and topic sentences, you may not realise until well into your essay if it's not working. By that stage, you are already fully committed to your argument and it's very hard to change track.

Planning time can seem like a luxury when you only have 40 minutes to write an essay, but I believe it's an absolute necessity. Taking a minute or two to break down the question and key words, write a skeleton thesis, and ensure that the topic sentences align, will save you time. If you have a clear structure and direction for your argument and you are confident you can sustain it throughout your paragraphs, you will think and write with increased speed, clarity, and flow. Having a pre-planned master essay or scaffold makes this process even easier. In the absence of a master, planning before writing in an exam is even more vital for the creation of a cohesive argument.

Essentially, planning an essay, just like planning anything, costs a little time upfront, saving a lot of time later, and is thus instrumental in academic success.

Side note: although I have specifically put this tip in the English section, the same principle applies to any extended response or essay regardless of the subject. A well-planned response will result in better quality and quicker execution.

TIP

Look for the marking criteria in every exam – if it exists, be sure to find it!

I mentioned this previously in the marking criteria section, but it's so vital that I want to reiterate it here. There are times, such as the final HSC exam, where the marking criteria may be separate to the question. Pay attention and look out for any criteria as you read through the exam paper. It can be easy to miss, and if the question is answered without regard for the marking criteria or the instructions of 'what you will be assessed on', you will likely compromise your results. Sometimes the directive term is placed in the criteria rather than the question itself, so neglecting this can be very problematic.

TIP

Including evaluative language in your analytical essays will benefit you.

This relates to the previous tip regarding seeking out marking criteria and directive terms. It never hurts to sprinkle some evaluative language throughout your essays, using words such as 'effectively' or 'skilfully'. If the rubric demands evaluation, you automatically tick the box. If you have missed an evaluative command, you have covered yourself. And if the rubric doesn't demand an evaluation, you have done no harm, and it still elevates your expression.

***

In this section we have covered a tremendous range of tips and techniques for excelling in English, and applying these strategies should provide a significant academic edge. Unlike subjects such as maths or science, which require a stepwise approach with each year relying on knowledge acquired in the previous year, English requirements do not change substantially each year. The skills learnt in Year 7 remain relevant all the way to HSC, and progression comes through the elevation and refinement of the same fundamental skill set. There are a few exceptions, such as strategies specifically geared to the HSC, and where applicable I have pointed these out.

My overarching advice for English, is therefore to take it seriously and start building the skills from Year 7. English is compulsory and can be stressful, but working smart, preparing well, and employing a strategic approach will set you up for success.

7. Maths strategies for assessment preparation and exam techniques

More than 75% of HSC students are usually enrolled in a mathematics course in NSW. Maths is therefore relevant for most high school students, so it's worth sharing some valuable tips for preparation and exam techniques.

Personally, mathematics was integral to our high school journey as both kids studied 4 Unit Maths (Extension 1 and 2) for HSC. Maths can be challenging because the workload is heavy, the content gets increasingly difficult and skills are built over many years. Equally it can be very rewarding for those who approach it with the right mindset. Maths is first and foremost reliant on two things: **consistent practice** and **perseverance**. The only way to excel in maths is to do a lot of it, and to be persistent in the face of challenge. With that in mind, let's look at some strategies for exam preparation and exam techniques …

TIP

Keep up with the work!

While it's important to keep up with all your subjects, it's especially crucial for maths, being a subject which builds upon itself quickly, with each concept reliant on a concrete grasp of the one before. Falling behind makes it hard to proceed with confidence.

Maths almost universally entails a lot of work, with most teachers assigning homework every lesson. To succeed, it's imperative to stay up to date with all classwork and homework each day, which will accumulate to many hours over time. Therein lies the foundation of your consistent practice and ongoing exam preparation.

Staying up to date also means by the time you have an assessment, all the basic work will be complete, and your conceptual understanding will be well established, allowing you to focus on higher level revision and practice.

TIP

Make very brief notes.

The main things to make notes on:

- Key concepts, rules, or formulae
- Tricky questions (with worked solutions)
- Alternative methods
- Traps and mistakes

This is something I have already mentioned, and it's a particularly valuable strategy for maths. I recommend having two separate notebooks at home:

- One to note key concepts, rules, formulae, tricky questions, and alternative methods.
- One to record error patterns, traps you might fall into, or recurring mistakes (even if they are simple and careless).

The things you choose to write in these notebooks could originate from classwork, homework, a textbook, practice papers, or actual exams. The aim is to record these in an organised way, using them as a study resource which you can review or redo before an exam. By reminding yourself to watch out for the traps and mistakes and refreshing your memory on rules, difficult questions, or alternative methods, you create an opportunity to elevate your exam performance.

> **TIP**
>
> **For exam revision, past papers are your best friend!**

Past papers are essential for maths exam preparation and neglecting this will almost definitely compromise your results. Although brief notes are helpful, especially the ones mentioned in the previous tip, success in maths is less dependent on passive revision and heavily dependent on **practice**. Past papers are not the only practice resource; there is also value in doing textbook exercises, chapter reviews, and mixed revision sourced from elsewhere, but if you had limited time and could do only ONE thing, it should be the most recent past paper (ideally from the same exam in the previous year). Most schools provide past papers and solutions for students to work through in the lead-up to an exam. For the HSC, past papers and worked solutions are available on the NESA website and these are essential to your HSC exam preparation.

When my son was in Year 7, his first maths assessment notification was accompanied by seven past papers. Being new to the assessment arena, I asked a friend with a son in the year above how many past papers she thought it was necessary to do. She looked confused by my question and simply replied: "Do all of them." At the time I thought that seemed a bit excessive, but my son followed the advice and did all the papers. The value of those past papers became clear very quickly, and from that point onward past papers were the top priority for every single exam … not just for maths, but ALL subjects.

We have already elaborated on the benefits of past papers, but I will do a quick recap on past paper strategies because this is such a crucial part of success in maths exams:

- Do as many past papers as possible because it's the best form of practice.
- If possible, try to procure past papers from other schools too – especially in Year 12, or in the event of a recent syllabus change where there may be a shortage of relevant past papers from your own school.
- If time is tight, prioritise the most recent ones.
- Do them under strict timed conditions.
- Mark your papers straight after doing them.
- Look at full worked solutions, not just answers.
- Evaluate your performance noting any queries, recurring errors, traps, tricky questions, or alternative methods.
- Resolve any issues as soon as possible in advance of the exam.
- The day before the exam – review difficult questions, retry select questions you got wrong, and revise your notes and traps.

A few related TIPS

- **Plan to do the past papers well in advance of the exam because they take time.**
- **Try to avoid doing a past paper the night before an exam – if it raises questions or you don't know how to solve a problem, you don't have time to resolve it, which can undermine your confidence going into the exam.**

Past papers are a significant part of the most valuable 20% of work you do in the lead-up to any exam and will be outcome determinant. They should be prioritised, ideally falling into Quadrant 2 work, which is non-urgent quality preparation.

TIP

Include a wide variety of revision.

While past papers are essential exam preparation, students aiming to excel go beyond this, doing a greater volume and variety of revision to expand their repertoire.

- It's best to utilise a range of sources, with varying question types and difficulty to increase the breadth and depth of your knowledge.
- Try to find resources with questions across topics, rather than just within, because exam questions don't stay in silos.
- It's great to practise harder questions, especially those which require the application of a combination of concepts, topics, or techniques. These equip you more thoroughly for challenging exam questions.
- It's also necessary to practise easier questions to cement your foundational skills. These operations become automatic with practice, increasing speed and accuracy in the exam.
- If you find you are weaker in specific topics, try to load extra revision which will focus on building skills in that area.

***

The best preparation can come undone in the absence of good exam techniques, so here are some tips for navigating the exam itself …

TIP

Don't allow yourself to panic during exam 'reading time'.

At the start of an exam, especially in Years 11–12, you will have some 'reading time'. This allows 5–10 mins to look through the paper before you are permitted to write. It's not uncommon for students to spot difficult questions during reading time, and if they don't immediately know what to do, they can panic. The problem is that once the exam begins, they fixate on those hard questions, reducing focus on all the other questions that they **can** do. This is where careless mistakes creep in, because rather than thinking about the question they are currently working on, they are distracted, stressing about a future question. The irony is that by the time they eventually get to the dreaded question, it often isn't as daunting as it initially seemed, and in the meantime, they have squandered easy marks on the rest of the paper.

Therefore, if you spot a hard question during reading time, DO NOT PANIC! Simply remind yourself to focus on each question as you methodically work through the paper, and you will deal with the hard question when you get to it.

> **TIP**
> **Always show working in an exam – even for a 1-mark question.**

This was another hard-learned lesson! It is possible to get 0/1 with a correct answer, just because there is no working. So, even if the question is simple and easily solved with mental maths, it's still important to show at least one line of working (unless it's a multiple-choice question). As unfair as this seems, the marker wants to see evidence that students know how to work it out and can demonstrate problem-solving skills.

> **Further TIPS for showing working.**

- When deciding how much working is necessary, the mark allocation is a reasonable guide. A 4-mark question requires a lot more working than a 2-mark question.
- Try not to skip steps or do too much mental maths, as marks can be awarded for specific interim working. If required steps are not shown, you may not earn all the available marks.

- Skipping steps can also be an unfortunate source of careless mistakes. Writing out working helps reduce these errors and makes it easier to find and correct errors if you have time to review your work at the end.
- Sometimes marks can be awarded for 'error carrying forward', so if working is shown and the examiner can see exactly where a mistake occurred, you might still get marks for subsequent working.

TIP

Always attempt at least some working – even if you have no idea what to do!

Even when students think they have no clue how to solve a question, making a start and writing down a few lines of working can translate to unexpected marks, because you might know more than you think, or it might prompt an idea for what to do next, or you might just get lucky. If you have time, trying to go step by step can help you to figure it out. If you write something down you might get a couple of marks, but if you leave it blank you are guaranteed to get zero marks.

TIP

Allocate exam time according to marks, and don't dwell for too long if you are stuck.

Maths exams tend to be tight for time with a ratio of approximately 1.5–1.8 minutes per mark. The general advice is to spend a bit less time on the easier questions and those with lower mark allocations, allowing a bit more time for challenging questions or those with higher mark allocations.

Don't ever spend too much time on multiple-choice questions. If you are struggling with a multiple-choice question, make an educated guess, circle it, move on and come back to it later if time permits.

There is no point spending 10 minutes on a question worth 1 mark and forfeiting higher value questions later in the paper.

A similar rule applies to **all** questions – if you are stuck and spending too much time on a question, move on and come back to it later – beware of the 'sunk cost fallacy'.

Bonus TIPS

Questions don't always progress in order of increasing difficulty through the paper, or even within a multi-part question, so if you find a question or sub-question difficult, remember that the following questions may be more manageable. In a related point, sometimes 'part b' of a question does not depend on, or even relate to 'part a', so if you can't do 'part a', don't assume you won't be able to do 'part b'. It's sometimes possible to complete 'part b' and win marks even if you haven't attempted 'part a'.

If you feel stuck, it's advisable to move on, bank as many marks as possible, and attempt every question in the paper, before returning to incomplete questions. Another benefit to moving on is that often while you are busy solving other questions your brain keeps working in the background and by the time you come back to a question you were stuck on, you suddenly know what to do. This happens more often than you think!

TIP

When using a formula – always show the original formula before substituting in values.

This was another hard-learned lesson! When there is a formula, students often forget to write the formula out, instead skipping directly to substituting in the values. It's entirely possible to do a correct substitution, get a correct answer and lose a mark for failing to write the initial formula. This is one of those mistakes that should be recorded in the book of traps.

TIP

When working with equations, substitute your final answer back in to check it works.

Even if you are confident your working is correct, it's a good idea to quickly substitute your answer back into the original equation to check. If it doesn't work out, it's an indication you may have made an error.

TIP

Check your work at the end and apply a 'sanity' check.

If you get time at the end of an exam, go back and scan over the questions, checking your working and answers to catch any errors. Use every available second to check your work until the time runs out.

Additionally, it's advisable to do a 'sanity check' to ensure your answers are logical and sensible. For example, in a simple maths problem you might want to check the answer approximates what you would expect. You also want to check that your answer is actually possible, for instance, if paint comes in 1L drums, and you need 4.75L of paint, you will need to round up and buy five drums. Furthermore, it's not possible to get a negative length or a fraction of a person ☺.

TIP

Always have at least TWO approved calculators with you in an exam.

This is essential because if your calculator runs out of batteries during an exam, you will be completely stuck. Schools often recommend bringing spare batteries, but it's far easier to use your spare calculator than to change batteries during an exam.

<u>TIP</u>

Double check your calculator is in the correct mode during your exam!!!

I only make this point because it's another lesson we learnt the hard way.

Before attempting to solve a maths question, double check your calculator is in the correct mode for either degrees or radians. It's soul crushing when you know how to solve a question, but you have accidentally used the incorrect calculator mode.

BONUS <u>TIP</u>

If you have TWO calculators as suggested, keep one in radians and one in degrees. This keeps things simple and helps mitigate the risk of using the incorrect mode.

A few more reminders and <u>TIPS</u> for eliminating all those 'careless' mistakes

- Read the whole question properly and **highlight** or <u>underline</u> key words.
- Copy the question correctly and transcribe correctly within your working.
- Make sure your handwriting is legible for the marker and yourself.
- Practise basic arithmetic too – you are never too advanced to make sure you can add, subtract, multiply, and divide simple numbers accurately – this is often an infuriating source of careless errors.
- Check your working and answer is in the correct units, rounded to the correct number of decimal places, estimated to the correct number of significant figures, etc.

- Ensure that you have accounted for all aspects of the information provided in the question, including any ranges or limits.
- Once you have got your solution, have a **quick** look back at the question before moving on, because you may pick up something that you missed.

8. General exam techniques

In this chapter, we have so far looked at broadly applicable assessment preparation as well as subject-specific preparation and exam techniques for maths and English. The reason we devoted so much time to English and maths is because they are the two biggest subjects throughout high school and the HSC and both have unique requirements. However, many of the exam techniques outlined above are equally important and applicable to other subjects, so with that in mind, here are some **final tips on exam techniques** which should be applied to **ALL SUBJECTS** whenever possible.

TIPS for optimising exam 'reading time'

- Many exams in the senior years will have 5–10 minutes 'reading time' at the start (before you can begin writing). It's important to use this time strategically.
- Skim or skip multiple-choice questions during reading time and rather look over other questions first.
- Read the harder questions so your brain can get to work on them subconsciously.
- Identify the pattern of questions, note where the tricky or time-consuming ones may be, and get a sense for the distribution of marks throughout the paper.
- Use the information gleaned during reading time to help guide your time management strategies – for example, if you see that

the last question is worth nine marks, ensure you leave enough time to work on this.

- Do not dwell on any single question during reading time.
- If you encounter a question that concerns you, DO NOT PANIC! You'll deal with it when you get to it later. If you spend the exam fixating on a hard question, it causes distraction, potentially derailing you and costing marks in easier questions.
- For English and history: read the essay questions first, then the short answer/source questions, and then progress to the passages/sources/texts last.

TIPS for answering multiple-choice questions (MCQ)

- Maths, the sciences, economics, business studies, and various other subject exams start with a multiple-choice section which could contain up to 20 questions.
- Don't spend too much time on this section, or any specific question, as each one is only worth one mark.
- The ratio in most exams is 1.5–1.8 minutes per mark (depending on the subject, year level, exam time and marks available), but you should spend less than 1.5 minutes on each multiple-choice question as they are generally less complex, and marks are not awarded for working or explanation. Spending less than the allotted time on MCQ allows more time for complex questions later in the exam.
- Paradoxically, you might still encounter a very difficult multiple-choice question, and if it's taking too long, rather make an educated guess, flag it for review, move on and come back to it if you have time later.
- If you can't work out the answer, the next best method is to try eliminating at least two options and making an educated guess, giving you a 50% chance of being correct.
- If the four options are A, B, C, D, do not overthink the pattern of responses – just because you choose three consecutive option Cs, doesn't mean you are wrong.

- Make sure to fill in the option that corresponds with your intended answer – this sounds logical, but when rushing and stressed it's possible to get the right answer and accidentally select the incorrect option.
- Be aware of multiple-choice traps, where one of the options caters for an error commonly made by students.
- Double check the units in the answer options – often incorrect units are traps.

TIPS for exam time management and general strategy

- Effective time management is one of the most crucial exam techniques to master, as poor time management will cost you marks regardless of the subject.
- You should aim to attempt every question and finish the paper, preferably with a bit of time to spare so you can check your work.
- If there was reading time, you should have an idea where the time-consuming and high mark value questions will be and accommodate for this in your approach.
- We have already said that the time per mark is approximately 1.5–1.8 minutes per mark in most exams (especially the HSC), so try to complete easier questions in less than the allowed time, leaving more time for complex or lengthy questions ...
- ... BUT be careful not to spend **too** little time on 'easy' questions because rushing excessively increases your chances of making careless errors, and you don't want to lose easy marks.
- Similarly, don't confuse less time with less focus – treat easy questions with the same focus as harder questions, because if you do not pay attention, you will squander marks with careless errors.
- Manage time according to mark allocations – spend less time on questions worth less marks and devote more time to questions with large mark allocations (noting that mark allocation does not always align with difficulty level).

- If you get stuck on a question, be mindful of how long you are spending on it. If it's taking longer than it should, circle it so you know to come back to it and MOVE ON. As a rule, if you have tried to solve a question twice and you aren't getting it, move on and don't fall prey to the 'sunk cost fallacy'. There's no point spending twenty minutes on one question and running out of time to do the last three. It's better to move onto the remaining questions, returning later if there is time. The other benefit is that often when you come back to a question, you have greater clarity on how to solve or answer it.
- Remember that question difficulty does not always increase progressively through a paper, and parts of a question can be independent. Quite often there will be a challenging question followed by a more accessible question. This is another reason to stay calm and move on, giving yourself a better chance of winning marks.
- If you are running out of time, do the questions you KNOW you can get right first, bank some marks and then move onto harder ones.
- Remember that questions in an exam can often be completed in any order
    - It can be helpful to complete easier questions first to boost confidence …
    - … but be strategic in your approach, and mindful that jumping around can cost time and increase the chance of a question being skipped accidentally.
- In essay subjects, you will focus less on the minute per mark and more on managing the overall time. For example, if you have three essays to complete in 120 minutes (as per the English Advanced HSC), you need to be disciplined in your approach, ensuring you allow approximately 40 minutes per essay.
- Using diagrams and tables in your answer (rather than writing walls of text) can be an effective way to enhance clarity and save time, especially in the sciences.

- Underlining and circling words in your answers to make them stand out can be effective to add clarity for yourself and the marker – this can be done in a variety of subjects, including the sciences and humanities.
- Don't leave any questions blank – always write something to gain partial marks.

TIPS for reading your questions, instructions, and marking criteria properly

- Read each question properly all the way to the end.
- Pay attention to all parts of questions and sub-questions.
- Read and highlight all secondary instructions, including requirements to justify your response, provide examples, apply limits, or specifics on how your answer should be represented.
- Note directive terms and key words of the question and underline or highlight them.
- Break complex questions into manageable parts.
- Actively look for any marking or assessment criteria.

TIPS for planning extended responses and essays

- Highlight and chunk parts of the question, accounting for all parts in your answer.
- Take a minute or two to plan before you write. Jotting down a brief skeleton or key notes can help keep you on track and improve your flow.

TIPS for writing

- Ensure your handwriting is clear and legible regardless of subject – if they can't read it, they can't mark it!

- Use proper spelling, grammar, and punctuation – it creates a better impression.
- Use the correct technical terms when you write – this applies to all subjects.
- Short and longer answer questions usually require one point or piece of evidence per mark (for example, in science, English, history).
- Use the amount of writing space as a guide for how long your answer should be, and if there is extra space, use some of it (especially for English).
- Clear expression, precision, and accuracy are key – marks are awarded for **what** you say and **how** you say it – it's possible to lose marks on a question you know how to answer because you didn't express it in the right way. This is especially true in the sciences.
- Remember quality trumps sheer quantity – so know when you have written enough to answer the question, then STOP and move on to the next question.
- Take a moment to think through your response before you start writing.

TIPS for ensuring you don't miss questions.

- Check both sides of each exam page to make sure you haven't skipped questions accidentally. This is a common error, especially near the end of an exam, where students forget to **check the back page** and miss the last question/s.
- In general: don't assume the exam is finished until you see the words '**end of paper**' or '**end of assessment**'. Most exam papers will put this statement after the last question, so students know for certain that there are no further questions.
- Check the exam question numbers flow sequentially from one page to the next. If you are missing question numbers, check the pages aren't stuck together or missing, and if necessary, ask the invigilator.

TIPS for the end of the exam

- If you have time at the end of an exam, use every available second to go back and complete skipped questions and check your work.
- First attend to incomplete questions which you flagged and try to answer them.
- Then re-read questions to ensure you didn't miss instructions or directive verbs – this is quicker and easier if key terms or instructions have been highlighted.
- Check as many of your answers as possible, prioritising those you were unsure of.
- Check all your handwriting is legible and clarify any illegible words or workings.

***

If you have studied and prepared well, and managed to effectively implement these exam techniques, you are in a great position to excel. There is, however, one last piece of the performance puzzle that is often overlooked … it's the bit that comes after the exam! Once an exam is done, it's tempting to put it behind you and move on, but there are a few things that should be done to facilitate greater academic success. Let's look at some post-exam strategies which will assist you in preparation for future assessments.

9. After the exam

TIP

Avoid the exam post mortem and assumptions about your performance.

Immediately after an exam, students will compare answers, reinterpret questions and often work themselves into a panic, making assumptions about marks they may have lost and how this might impact their results. Even if students aren't having these conversations with others, many fall into the trap of replaying the exam in their mind.

My advice is to avoid this as much as possible because:

- It's not helpful – you can't go back in time and change your answers.
- It increases stress and promotes self-doubt.
- If your answer differs from your peers, it doesn't mean you got it wrong – perhaps they did!
- Quite often your assumptions are completely unfounded – you can think you did terribly and may end up doing brilliantly.

Obsessing and jumping to conclusions about your performance before getting the results is unhelpful and counterproductive, consuming valuable mental energy that could be put to better use.

***

Within a couple of weeks of completing an exam you will eventually receive your results. Hopefully you applied the strategies we have covered so far, and these have paid off, but regardless of whether the results are excellent, mediocre, or disappointing, there is still work to be done. Here are some tips and guidelines for what to do next…

TIP

When you receive your marked exam paper back … CHECK your marks thoroughly!!

Specifically check that:

- The marks are correctly added
- Every question has been marked
- You understand why marks were deducted.

The reality is that teachers are human and can also make mistakes. We experienced quite a few occasions where marks were incorrectly added, with the total lower than it should have been. There are also sometimes inconsistencies in marking, meaning students may be marked differently for the same answer.

If you don't understand why you lost marks, you may wish to request a re-mark. In any event, if you have concerns or queries, you should ask the teacher, because quite often the issue can be easily remedied, especially when marks have been added incorrectly. You shouldn't be afraid to ask – you have nothing to lose. Even if your mark doesn't change, the clarification provided by the teacher still provides a learning opportunity.

TIP

If you want to query a mark, do it immediately ... before leaving the classroom!!

Once you leave the classroom with your marked paper in hand, it cannot be queried later. For example, if you get home and realise the marks have been added incorrectly or you would like to ask for a marking review, it's too late. This rule is designed to keep things fair for all students, because once a paper has left the class, teachers can't be sure whether changes were made retrospectively.

***

A side note on checking marks:

Keep a record of your raw mark, percentage, and weighted contribution for each assessment task. For each assessment you will usually receive your raw mark and the calculated percentage. If you apply the published assessment weighting, you can calculate the weighted contribution to your year mark. For example, you might get a raw mark of 18/20 on an exam, which equates to 90%, and if the task weighting was 10%, this would contribute 9% to your year mark for that subject. Once you have completed all assessments for the year, you can easily add these to determine your final percentage for the year for each subject.

I suggest keeping a written record of these marks and calculating the annual weighted totals which can be checked against the final marks on school reports, because sometimes mistakes are inadvertently made. We personally experienced a few instances of marks being incorrectly recorded in the teacher's mark book, or incorrect assessment weightings being applied, but these errors were easy to identify, and once the school was alerted, they were easily remedied. Without this reference, it would have been much more difficult to pick up discrepancies, and small errors can have an impact on final grades, ranking, and even the achievement of academic awards.

A further benefit of tracking your cumulative marks in each subject is that it gives you a concrete way to evaluate your progress in the achievement of your academic goals. If you are aiming for a final mark of 94% in maths, and you can see that you have already achieved an average result of 95%, having completed 80% of the year's contribution, then you are likely on track to achieve your goal, needing to get 90% on your final exam. Conversely, if you are lagging behind in your cumulative average, you will need to overperform in the final to bring your mark up at the end.

TIP

Try to interpret your assessment and exam results in relative rather than absolute terms.

The absolute assessment mark often generates an instinctive reaction from students (and possibly parents), but it's important to remember that marks can only be properly interpreted in relative terms. For example, a result of 75% on a chemistry exam may be initially disappointing, but if the top mark was 76%, and the average was 60%, this creates an entirely different perspective, and 75% would be interpreted as an excellent result. Conversely, a result of 92% might initially seem amazing, but if the exam was deemed too easy, with a top mark of 100% and a cohort average of 94%, then suddenly 92% doesn't seem quite as brilliant. This is a difficult thing to come to terms with, and it can be tricky to navigate if there is a time delay between receiving the absolute mark and the other assessment statistics, which is often the case. It's therefore important to remind yourself to consider the relative distribution of marks before reacting prematurely or jumping to conclusions about your performance. This can be especially challenging for high-achieving students who are more prone to judging their performance in harsh absolutes.

TIP

Try to maintain a healthy perspective when it comes to ranking … and remember that collaboration is equally crucial.

Another metric of relative performance is student rank, which is used in Year 12 (and earlier in some schools). This is the ultimate comparative metric and tends to be a bit controversial because there are benefits and costs associated with this measure. If you are ranked first, this is fabulous, and a wonderful confidence boost, but it simultaneously adds extra stress and pressure to maintain the top spot. If you are ranked last, this can feel miserable and negatively impact your confidence, also adding stress, and often diminishing motivation. Every other point on the continuum comes with its own issues.

Ultimately, rank does matter! In the HSC, your rank order, based on your final internal cumulative school assessment results, will in most cases determine 50% of your final HSC mark. Students are acutely aware of this, and the significance of rank order can therefore generate intense competition, especially in the lead-up to trial exams, as each student wants to secure the best possible rank heading into finals.

While rank is important, and there is nothing wrong with a bit of healthy competition, students should try to keep a balanced perspective, because we have learnt that in most cases a few ranks difference has little material impact on the final result, and collaboration between students is equally, if not more important!

There can be a tendency for students to compete in the lead-up to trial exams, with a single-minded focus on their individual rank. After trials, there is a noticeable shift to information sharing, teamwork, and collaboration. Students understand that while part of the HSC result is determined by individual performance, the rest is determined by the cohort performance. The stronger the performance of the overall cohort, the better students within it will fare. It's therefore in everyone's best interests to work together. In addition to the academic benefit derived from collaborative efforts, it builds a sense of community and camaraderie, generating a positive, supportive environment.

TIP

Do not let a disappointing result get you down.

There is one guarantee in high school – no student has a linear trajectory to excellence. There are sure to be a few academic disappointments, and even the highest achievers have ups and downs.

Even in Year 12 when 'everything counts', the odd disappointing mark will not have any significant impact on your end results.

Remember:

- A poor mark does not define you
- There is always a learning – if you are willing to take it on board, you will improve
- Perseverance pays off
- Practice brings rewards …
- … so, keep working at it

Note for students and parents: I am not saying (and nor should you) that a disappointing result doesn't matter, because that can invalidate the hard work that was done, undermining the importance of effort and academic goals. The reality is that a disappointing result will often matter a lot, but if students can keep it in perspective and learn from it, then that one result will not define you or your ultimate grades.

TIP

Reflect on your exam performance critically and diagnostically – even if you did well.

After every exam, the best thing you can do is go through your paper in detail to see where you made mistakes, why they occurred, and identify broader error patterns.

Analysing your mistakes is a very important part of the learning process and growth mindset. If you can't diagnose where you went wrong and the reason why, you won't be able to remedy it and apply the knowledge to future assessments.

For example:

- If your mistake was due to lack of conceptual understanding, the solution lies in seeking clarification from your teacher, reviewing your notes and filling in the gaps.
- If your error was part of a pattern, because it's the third time you forgot to reverse an inequality sign when multiplying both sides by a negative number, then this should be added to your book of **traps** to ensure you revise it before your next exam.
- If your mistake was not addressing the directive term, the solution lies in revising the directive term definitions and taking a highlighter to your next exam, ensuring you highlight the relevant term in each question so you **can't** overlook it.

Once you have been through your marked paper, it's advisable to review the marking criteria and model answers if you have them. It's one thing to consider your results in isolation, but even better to take a bigger picture approach, understanding what markers were looking for when awarding marks for each question. This provides a clearer roadmap for improving future performance. It's important to fully understand why each mark was lost and ensure that all errors are clearly resolved. If there is any uncertainty, it's best to seek clarification from your teacher promptly, so you avoid a situation where gaps in understanding remain unresolved, impacting future assessments.

As you go through this process, take the time to add to your notes and your book of traps and mistakes to enhance your revision resources for the next exam, thereby reducing your chances of repeating mistakes or misunderstandings. While most of this reflection tends to be focused on mistakes and improvements, it's equally important to recognise what you did right, what earned you 'ticks', and what aligned with the expectations of the marking guidelines, so you know what to replicate in future.

This process of review, reflection, and refinement of skills does take a bit of time, so it's often neglected and treated as a low priority, but it's an extremely valuable exercise which pays dividends for students who make the effort. It's a Quadrant 2 activity, which also contributes to the most valuable 20% of your work.

TIP

Take on board ALL the feedback.

Beyond analysis of what you got right and wrong in individual questions, there is a highly valuable opportunity after every exam to reflect on broader feedback. This is often given from a macro perspective, so it's important to incorporate this into your improvement plan for the next assessment. This advice is relevant regardless of whether your result was good or not. You are never truly finished improving. Teachers recognise this and want to give you the tools to take your work to the next level. Some students find feedback demoralising, and in fact it took us a while to realise that detailed feedback did not necessarily indicate the work was lacking. Often this feedback was designed to help transform great work into excellent work.

Bonus TIP for Year 12

For subjects where prepared materials will be used again in a similar form in future assessments, such as English master essays or history essay scaffolds, you should rework these immediately to incorporate the feedback. In Year 12, you will be examined on the same content multiple times throughout the year, so if you are diligent about this process, you will always be working from the best version of your materials. Conversely, if you don't make these improvements timeously, you risk forgetting or running out of time to make the adjustments and may end up repeating the same mistakes.

10. A note on the additional demands of Years 11–12

The skills and techniques we have covered so far in this chapter apply to all years from 7–12 (unless specified otherwise). Good preparation and exam techniques will undeniably pay off, creating academic benefit right from Year 7. However, as you progress through Years 11–12, the demands increase, available time decreases, and skills that were advantageous early on become critical for academic success. The optimal approach is consistent application and skill building from Year 7, so by the time you enter Stage 6, these skills are well entrenched. However, if you have chosen to take a more relaxed approach, it's important to pivot your methods if you are to achieve your academic goals in the HSC.

For instance, in Year 7, it's beneficial to proactively make study notes as you go, but if you can't manage this, you will usually still have time to prepare before an exam, and if you are effective and efficient it shouldn't compromise your results much. However, in Year 12 if you receive your trial exam notifications and you haven't made your study notes yet, this is going to be very problematic. With this in mind, let's look at a few of the most significant changes in Years 11 and 12 …

The jump from Year 10 to Year 11 is enormous:

- Year 11 is only three terms, which is incredibly challenging – you are still required to complete four terms worth of work, meaning there is a lot more work and a lot less time, so a much faster pace.
- There are only three assessments per subject, with each one having greater weight.
- The volume and complexity of work for each subject increases dramatically.
- The difficulty of assessments increases substantially.

- Marking becomes stricter and it's harder to earn very high marks (do not be alarmed if your marks drop a bit).
- Year 11 final exams (Prelims) are mostly on the full year of work.
- Year 11 Prelim exams are longer.
- Year 11 results count and are used for early entry university applications in Year 12.
- Year 11 subjects mostly carry through to Year 12.
- Year 11 subject content becomes assumed knowledge in many Year 12 courses, which means you may be examined on elements of it in Year 12 exams (depending on the subject, but certainly for maths).
- Ranking comes into play formally in many schools, adding pressure and stress.

The jump from Year 11 to Year 12 is slightly more moderate:

- In Term 4 of Year 11, you begin Year 12.
- From the first day of Year 12, every result counts.
- Everything you learn from the first day is examinable in your finals (in addition to assumed knowledge from Year 11 if applicable).
- You may or may not choose to drop one of your Year 11 subjects (you must do at least 10 units for the HSC, 2 of which must be English).
- Trial exams and final exams are very long (many will be approximately three hours).
- You may be juggling extra formal responsibilities if you hold a leadership position.
- It's a physically, emotionally, and academically demanding year, spanning almost five terms.

With the added workload and pressure, there isn't much time to experiment, and those who have used Years 7–10 to develop their processes and skills will benefit significantly. For those who haven't managed this yet, Year 11 is the next best time to start.

If I could distil my advice into one key tip, it would be this …

TIP

Be PROACTIVE, not reactive!!!!!!!!

The single most beneficial change you can make from the start of Year 11 is adopting a proactive approach to assessment preparation, where you anticipate and work ahead, rather than waiting to be told.

In our experience, the biggest area of impact is **making notes proactively**.

We have established that making notes is the first step in most assessment preparation, and for many subjects it's also one of the most time-consuming activities. Students who have kept up to date with notes from the start, will find it much easier to manage the workload as it builds. It also creates a crucial academic advantage, because when exam notifications are issued, much of the foundational work will already be done, and preparation time can be devoted to study, revision, and practice.

Students who have not kept up to date with notes will find themselves in a predicament, because the limited time leading up to exams will need to be spent making notes, which compromises the overall quality of preparation. Either notes will be rushed to accommodate study and practice, or taking time to make good quality notes will reduce the time available for study and practice. Neither of these scenarios is optimal, but with a reactive approach, this trade-off is inevitable. From Year 11 onwards most assessments take place in exam blocks, which means you will receive multiple notifications at a similar time, and exams will likely all be done within a week or two, so if you are in the same tricky spot with all your subjects, the problem will be amplified, and so will the stress, as you operate primarily from Quadrant 1.

It's incredibly difficult to stay completely up to date with notes, especially for content heavy subjects. However, even thirty minutes of notes per weekday evening, if done consistently, will generate gains, because this accumulates to 2.5 hours a week. That amounts to an extra twenty-five hours over a 10-week term, which would be hard to find otherwise. If you can aim to do thirty minutes of notes a night during the week, a bit more on weekends, and be no more than a week behind on your notes for all subjects, you will be very well placed to manage whatever is thrown at you throughout the term.

Making notes is just one aspect of exam preparation, but the same principle applies equally to other preparation activities. If you know you will need to do it, don't wait to be told – just get it done.

A second area of impact is **starting well before deadlines**. This is a good strategy to employ generally, but especially in Years 11–12 when workloads are immense. The earlier you can get ahead on a task, project, or your broader exam prep, the more manageable your load will be. This is also a great way to ensure you have a buffer to accommodate for any potential curveballs.

As I have mentioned a few times already, it's imperative to **stay up to date with all classwork and homework**. This is important in every year, but falling behind in Years 11–12 will be extremely problematic. Once you have fallen behind, it's very challenging to catch up, given the volume of work and the pace of the Stage 6 curriculum. Furthermore, if you are behind in these fundamental areas, it impedes your capacity for broader assessment preparation.

Whether it's classwork, homework, notes, or revision exercises, it's strategically advantageous to tell yourself you are preparing for assessments from the first day of term, because easing into it for even a few weeks will ultimately result in falling behind. This isn't easy, and requires all sorts of self-discipline, but for those aiming to achieve at the highest levels, it paves the way. Working proactively within Quadrant 2 creates significant value, helping to generate the results you want.

> **TIP**
>
> **Do some work in school holidays – but also pace yourself if you can.**

Some school holidays are pure holiday, and some might not be. Once you start Year 11, holidays become more like study breaks. The amount of holiday time devoted to studying depends on the school, the subjects, and the student's academic goals.

Some advice for school holidays:

- If schools assign holiday homework, it's advisable to do it, because by Stage 6 it's generally necessary for your success.
- Use the time to catch up on any incomplete classwork and homework.
- Do all your notes up to date for all subjects so you start the term on the front foot – once term begins it can be difficult to carve out time to catch up on notes.
- If you have assessment notifications, work diligently to prepare as much as you possibly can, because it's harder to find the time once the new term starts.
- If you have an English set work, read it!
- If you are in a position to draft a master essay, holidays are a great time to do this (especially in Year 12).
- It's advisable to rest a bit in the September break after Prelims, before Year 12 starts.
- Once Year 12 starts, most of the holidays, including the long December break, will include some academic preparation.

- In the long December holiday after the first term of Year 12, try to compartmentalise parts of the holiday that you will use for study, while ensuring you also take a break, because once the new year starts, there will be very little time to rest.

***

In this chapter we have covered a variety of preparation strategies, exam techniques, and general tips to assist you in acing your assessments all the way from Year 7 until the end of HSC. By now you should have a solid grasp on how to navigate the assessment system to achieve the best possible academic outcomes. There is, however, one more vital strategy needed to clear the path to success … and that is the all-important planning.

PART 3:
STRATEGIC PLANNING FOR ACADEMIC EXCELLENCE

"Always think of progress in small increments" (D.W. – 99.0 ATAR)

CHAPTER 9:
Assessment and study planning

At the beginning of this book, I alluded to the importance of planning, and I firmly believe this is the cornerstone of academic achievement. The best intentions can quickly unravel in the absence of planning, while great planning allows for strategic utilisation of available resources, creating optimal conditions for academic success.

When it comes to managing assessments, planning is particularly crucial because you will often be working on a few (or all of them) at the same time, with tight deadlines, other schoolwork and personal commitments to juggle, and significant time constraints. Although some class time might be allocated to students for structured preparation in the early years, most assessment work from Year 7 onward is expected to be done independently outside school hours. It's therefore vital to implement the principles outlined in previous chapters, **planning** to ensure you work smart and optimise your resources, ideally maximising proactive time in Quadrant 2, and focusing on the most valuable 20% of work which will generate 80% of your results. The sooner you develop a system for assessment planning, the better off you will be, because you need to employ this throughout high school right up until the very last HSC exam.

Note: This section primarily deals with **assessment planning**, but the same principles should be applied to managing your daily workload beyond assessments.

Often students under pressure believe that they don't have time for planning because it takes time away from working on assessments. However, planning itself is an essential Quadrant 2 activity and part of the most valuable 20% of effort. If done properly, it saves time and generates better results. Lack of planning can result in poor organisation and poor time management, leading to increased stress and suboptimal outcomes, especially if students end up crisis-managing in Quadrant 1 and running out of time or forgetting crucial aspects of preparation.

1. Initial planning steps

In an earlier chapter we talked about the benefits of using a diary, creating a study planner, and developing an organised system to manage your daily workload. We also discussed the principles of prioritising, time management and the 80/20 rule. Effective assessment planning demands an extension of these fundamental frameworks, but there are a few additional tips and strategies which we will share in this chapter, which are tremendously helpful in managing a heavy assessment load.

TIP

Transfer your annual assessment schedule into your school diary the day you receive it.

At the start of every academic year (usually by Week 4 of Term 1), most schools issue an assessment outline for the year. This usually provides the following information:

- The number of assessments per subject for the year
- The terms and weeks in which each assessment will take place

- The weighting of each assessment task
- A provisional description of each task (written exam, hand-in essay, etc.)
- The dates of any exam blocks

It's advisable to immediately write every assessment, including the type of task and its weighting, into the relevant week in your diary, year planner, or calendar so you can start planning around it. This overview, which literally takes 10 minutes, will highlight crunch times and reveal the gaps where you can plan to devote more time to Quadrant 2 activities. The nature of the task and the weightings also provide an early indication of the type of work entailed and relative importance of each task. While details may still be vague, this preliminary information gives structure to your assessment workload, helping to guide your planning and prioritising holistically through the year. As details become clear later, you can update your diary and adjust your plans accordingly.

For instance, your Term 1 assessments might be spread out with two assessments per week for four weeks, meaning you will likely work on 2–4 assessments at a time, taking a steady and sequential approach over a 6-week period. However, in Term 4 you may have an exam block with eight exams in ten days, which will require you to start ahead of time and work on all eight assessments simultaneously. These two scenarios would require different plans to ensure that you can get all your assessment preparation done in time.

Note: Schools may schedule exams differently in Years 7–10, but in Years 11 and 12 most exams, including trials and final HSC exams, take place in exam blocks with all exams occurring within a few short weeks, leaving little breathing space. Therefore, by the time you get to the final stages of high school, the exam timetables are gruelling and if you have not implemented streamlined scheduling and planning systems, you might find yourself in trouble.

TIP

Plan around your assessment schedule as much as possible.

Once you have scheduled your assessments into your diary or planner, try to organise the rest of your commitments around this as much as possible. For example, if you know you have an exam block in Week 6, perhaps don't book a social weekend away in Week 5. If you can schedule appointments and social commitments into weeks where you are under less pressure, clearing as much space as you can in the lead-up to big assessments, it will help to free up study time and alleviate some of the stress.

TIP(S)

When you receive an assessment notification:

- **First: write the specific due date or deadline into your diary/ planner**
- **Second: read the notification thoroughly to understand the requirements of the task**
- **Third: make a detailed to-do list and highlight high priority items.**

Most assessment notifications are issued two weeks prior to an assessment – a bit longer if you are lucky – which means that time is often tight from the start. A methodical approach will therefore serve you well.

First: Write the exact due date into your diary, replacing the provisional timing recorded from the annual assessment guide. This ensures you will not forget and gives you a sense of your lead time for preparation.

Second: We established previously that it's vital to go through the assessment notification carefully, ensuring you have a good grasp of the requirements. This entails a detailed understanding of **what you need to do, what you need to know, what questions you will need to answer** and **what the marking criteria** are.

Third: Once you know what you are working towards, the next step is breaking the overall task into a comprehensive list of subtasks, which can be individually prioritised and planned into your diary or study planner. It's very hard to plan effectively for something as broad as 'study for science theory exam' or 'prepare for unseen essay exam'. As such, the best quality and most achievable plans are those which are broken down into specific components which you can work through systematically. As you break your tasks into smaller components, try and work to the lowest practical level of detail, so each subtask is as defined and specific as possible.

Deconstructing bigger tasks into smaller parts is beneficial because:

- It helps you identify (and remember) all the discrete steps you must take to prepare properly for the task.
- It's easier to plan when you have laid out the full scope of the task.
- It illuminates high priority items.
- It makes your plan specific and actionable – vague plans are harder to implement.
- It's easier to plan time for a great number of smaller tasks than a small number of greater tasks.
- Onerous tasks seem more manageable when broken into digestible chunks – this in turn reduces the potential for procrastination, which can result from students feeling overwhelmed and not knowing where to start.
- It helps you track progress and feel more in control.
- Completion of subtasks creates a sense of accomplishment.

Once you have your list of subtasks, it's good to highlight any obvious priorities, as this helps guide your plan later. If you have multiple assessment notifications this is even more important because you will need to make time for all the subtasks across multiple lists, with consideration for priority both **within** and **between** assessment tasks.

TIP

Next, assign approximate timings to each subtask and calculate the total time.

Before you can create an effective study plan, you must estimate how long each subtask on your preparation list might take to complete, so you can allocate appropriate time. If you schedule tasks without considering their timing, you could end up with a plan that is too ambitious and difficult to fulfil, which can be frustrating, demoralising, and stressful. Alternatively, your plan might underutilise your capacity, spacing things out so much that you run out of time at the end, which is equally frustrating, demoralising, and stressful. Neither of these scenarios is ideal, and good planning requires **proper time management** to ensure you create a balanced, achievable, sustainable schedule with consistent workflow, avoiding too many peaks and valleys.

The trick is to be **realistic**, and not overly optimistic when assigning time to preparation tasks. It's not helpful if you allocate too little time to items on your list, because if you can't complete them, it sets you back and creates additional stress. For example, if a maths past paper stipulates an exam time of two hours, then estimating only two hrs to complete the task, allowing no additional time for marking, would be unrealistic. Similarly, estimating two hours to make all your science notes would likely be too ambitious and probably unachievable. Initially, time estimates may involve some trial and error, but as you progress further into high school you will get a feel for how long certain tasks usually take you. If anything, rather err on the side of generosity, giving yourself a time buffer to ensure you can complete your planned tasks.

***

Regarding these last two tips – here are a couple of real-life examples of assessment preparation to-do lists and accompanying time estimates…

Maths assessment preparation list (for a written exam):

- Complete classwork – Questions 3, 4 *(15 min)*
- Complete homework – Exercise 6E and mark *(1 hr 15 min)*
- Make notes for textbook Chapters 4, 5, 6 *(20 min x 3 = 1 hr)*
- Do textbook chapter reviews for Chapters 4, 5, 6 and mark *(1 hr x 3 = 3 hrs)*
- Do relevant past papers plus marking: 2022, 2023, 2024, 2025 *(1.5 hrs x 4 = 6 hrs)*
- Do revision sheets given by the teacher 1, 2, 3, 4 and mark *(1 hr x 4 = 4hrs)*
- Review class topic tests *(1 hr)*
- Review previous relevant assessment task(s) *(1 hr)*
- Retry difficult questions or previous mistakes *(2 hrs)*
- Revise notes *(15 min)*
- Revise traps *(15 min)*
- Review assessment notification and prepare equipment (*5 mins*)
- **TOTAL** time required: approx. 20 hours

Science assessment preparation list (for a written theory exam):

- Complete all relevant classwork *(10 min)*
- Complete all homework – Worksheet 6, Questions 7–10 *(20 min)*
- Make notes for each checklist item/syllabus dot point *(6.5 hours total)*
- Revise notes x 4 times *(1.5 hr x 4 = 6 hrs)*
- Do past papers plus marking: 2024 and 2025 *(1.5 hrs x 2 = 3 hrs)*
- Do textbook reviews for relevant chapters and mark *(2 hrs)*
- Do revision provided by teacher and mark *(2 hrs)*
- Review class topic tests *(30 mins)*
- Retry difficult questions or mistakes from revision and past papers *(1 hour)*

- Review assessment notification and prepare equipment *(5 mins)*
- **TOTAL** time required: approx. 21.5 hrs

Note: These maths and science lists are reflective of the earlier years of high school, and the timings would be extended for Years 10–12 as the work becomes more complex and time consuming, and past papers are longer.

English in-class essay exam (unseen question) preparation list:

- Read the set work *(done)*
- Highlight quotes and annotate techniques and analysis in the text *(in class time)*
- Review the NESA focus area statement for key concepts *(15 mins)*
- Make brief notes on author purpose, context, values, key message, ideas/themes, form, characters *(1.5 hrs)*
- Finalise quote bank *(2 hrs)*
- Review marking or assessment criteria *(15 min)*
- Create master essay plan *(2 hr total)*
    - o Craft master thesis *[30 min]*
    - o Craft master topic sentences for each paragraph *[30 min]*
    - o Select evidence from quote bank for each body paragraph *[1 hour]*
- Write first full draft of master essay approx. 800–1,000 words *(2 hrs)*
- Submit for feedback *(10 mins)*
- Finalise master essay implementing feedback *(1 hour)*
- Memorise master essay over a few days *(15 min x 10 = 2.5 hr)*
- Write full practice essays (adapting to Q2 and Q3) *(40 min x 2 = 1 hr 20 min)*
- Brainstorm skeleton adaptations (using past questions) *(15 min x 6 = 1.5 hrs)*
- Revise notes, quote bank, marking criteria, essay the day before the exam *(1 hour)*

- Review assessment notification and prepare equipment *(5 mins)*
- **TOTAL** time required: approx. 15.5 hours

A few things to note:

- While these are examples of real lists and timings constructed and used for exam preparation, list items and associated timings would vary for each individual student depending on the year, subject, nature of the assessment, amount of content, speed at which a student works, individual resources, goals, priorities, and preferences.
- The subtasks and timings are specific rather than broad or general.
- You will notice that the emphasis varies from subject to subject and depends on the nature of the task. For example, a science theory exam requires significantly more time to be spent making and revising notes than either maths or English, whereas English requires more practice preparing and adapting a master response, and maths has a heavier emphasis on past papers and practice revision questions.
- Each assessment task has its own tailored task list, although there are quite a few items that are relevant for many assessments.
- The key subtasks on your to-do lists for certain subjects probably won't change too much over the years, but the time estimates and total time required for study and preparation increases substantially as the volume and difficulty of the work escalates. The higher the year group and the bigger the exam, the more time would be devoted to each item on the task list. In Years 11 and 12, for instance, you would have substantially more content of greater complexity, so naturally your notes would take much longer, as would your revision and study. You would also do a greater number of past papers and practices, each of which would take longer to do and mark as Year 12 exams tend to be 2–3 hours each, in contrast to shorter exams in earlier years. Furthermore, as you progress through to the end of Year 12, you are tested on cumulative

content, necessitating a significant escalation in study time towards trials and finals as you revise and practise a full year's worth of work.

***

Once you have completed your detailed task lists and timing estimates, the next step is to create the plan …

2. Creating your assessment study plan

The principles and process of creating and managing an assessment study plan are very similar to the weekly homework/study timetable outlined in Chapter 5. The key difference is that assessment study plans generally span a longer timeframe with greater volume and complexity of tasks.

Assessment study plans can be created in various ways depending on personal preference. Some students prefer using a paper diary or calendar, while others prefer electronic planners, spreadsheets, or apps. It doesn't matter what format you choose, but after trying a few methods, we found that an electronic planner was the most efficient because it was quick and easy to modify when necessary. This was especially helpful in the lead-up to HSC as plans needed to be regularly updated.

TIP

Don't delay!

Make your study plan as soon as you can. Once you have assessment notifications and you know the deadlines, it's best to plan immediately, with a view to starting your assessment preparation as soon as possible. Even if it feels like you have lots of time, it goes fast, work starts to pile up, and things can quickly get away from you. You also don't know what curveballs lie ahead that might hinder your progress later. Working well ahead of deadlines will always benefit you, and the more revision you do, the better you will perform.

TIP

First schedule in all your non-negotiable commitments.

During assessment time, life still happens. Part of planning is realistically accommodating for all your commitments. Whether this is a compulsory music lesson, sporting fixture, prefect meeting, tutoring session, dentist appointment, or family event – if it must be done, then it must go into the schedule. Once these have been put into your planner, you will know what time is available for scheduling your assessment preparation and study.

TIP

Allocate every item on your to-do list into a timeslot in your planner, while applying the principles of 'S.T.E.P.':

- **Specificity**
- **Time management**
- **Evaluating logic**
- **Priority**

I created this **'S.T.E.P.'** mnemonic because it's easy and intuitive to remember and implement – although I kindly request that you grant me creative licence to swap the order of the last two terms when I explain them ☺.

Before delving into these strategic considerations, I wanted to mention that we personally found it easier to plan backwards from the due date, rather than forwards from the present date, as it gave a better sense of timing and created an appropriate sense of urgency. It's not necessary to schedule work backwards, but we found it helpful. Regardless of whether you populate your study plan forwards or backwards, you should plan with a clear focus on the principles of **Specificity, Time management, Priority** and **Evaluating logic**. Let's look at these one by one …

Specificity:

When scheduling tasks into your plan, ensure you are **very specific** about your goals – exactly **what** you aim to do in that time AND **how long** you expect it to take.

Here are some examples of how specificity might look in a plan:

- Maths Extension 1 – Task 3, 2024 past paper + marking (2 hrs)
- Science – make notes for Topic 4, checklist items 6–13 (2.5 hrs)
- English – review and implement draft essay feedback (45 mins)
- History – source analysis Practice Questions 1–3 assigned by teacher (30 mins)

Specificity of your goal regarding both **task and time** makes your plan more organised, actionable, and achievable. It also helps you track progress and prioritise. Scheduling 'time' without task specificity, using broad terms such as 'maths study', is not helpful or productive because if you haven't thought about the outcome you wish to achieve, you can't evaluate whether you have achieved it. Furthermore, without a specific goal to keep you accountable, you can easily fritter away three hours accomplishing very little, while deluding yourself into believing you have done a solid study session. Even if you have mentally detailed a specific task to work on within that time, if it's not clearly stipulated, you land up with a plan that is vague and confusing. This makes it hard to prioritise and track progress and creates the potential to forget crucial aspects of preparation. If you are planning for multiple assessments, this compounds, creating significant issues which can compromise productivity and results. On the flip side, scheduling specific tasks without considering timing also has adverse implications, as your plans may end up either too ambitious or too slack.

Planning specific time for non-specific tasks or planning specific tasks with non-specific times are equally problematic, and it's important to **plan for both specific tasks and specific times**.

Time management:

When scheduling tasks, **time management** is another crucial factor. We have already talked about the importance of realistically estimating how long each subtask will take and specifying this time in your plan, but equally important is making sure that you are conscious of your **broader time availability and constraints** when planning.

It's critical to be realistic about what you can fit into your plan and achieve within a given timeframe. Optimistically over-scheduling your plan creates issues, setting you up for unnecessary struggle. For example, if a past paper takes three hours, it would be inadvisable to squeeze this into a gap of less than three hours. This example is clear, but a more common issue arises when students have bigger portions of time, allowing them to slot in a few tasks, and land up overestimating what they can fit in. For instance, if you get home from school sport at 6pm, and you usually go to bed at 11pm, you might perceive this as a 5-hour gap of time … but if you planned five hours of assessment preparation, you might find yourself in a crunch since you have not accounted for homework time, eating dinner, or anything else you need to do. This creates unnecessary stress because you will need to compromise on something – either neglecting the homework, not completing the assessment prep, or going to bed later, none of which are ideal, and all of which have adverse implications for the next day. Students often fall into the trap of over-scheduling, even on free days (such as weekends or holidays). On a free day, it may seem reasonable to plan in four tasks requiring an estimated eight hours of work between 8.30am and 6pm, but in our experience even the best plans will not always run perfectly to schedule, and anything that you can't complete must then get moved to another time. If this becomes a pattern, the backlog can be overwhelming, and you may run out of time to complete the required preparation.

Despite your best estimates, some tasks will take longer than planned, while others may go quicker. Hopefully over time these should balance out, giving you time to do everything, but this relies heavily on realistic scheduling and resisting the temptation to overcommit. It's also advisable to plan in a few time 'buffers' here and there to account for timing miscalculations and unexpected curveballs or complications.

Note:

- Exercise caution when scheduling assessment preparation and study for Monday–Thursday, because with school, unpredictable homework, tutoring, and other commitments, there is often very little time left for assessment work. We tried to rather load assessment prep into weekends, and any additional work that could be accommodated on the busy weekdays was a bonus.
- Although I mentioned in one of the examples that scheduling eight hours of work on a free day may be ambitious, this depends very much on the stage, the student, and the goals. There are times when students may well be studying a lot more than eight hours a day in the lead-up to big exam blocks, especially in Year 12.

A final note on time management:

When creating study plans, one of the decisions you will make is the length of your study sessions. As discussed in an earlier chapter, there are benefits to shorter study sessions in terms of enhanced focus and reduced mental fatigue, but there are equal benefits to longer study sessions in achieving flow and building sustained focus. The ideal time structure of your plan will be driven by personal preference, but by Year 12 it's advisable to schedule in some longer sessions of 3+ hours to train yourself for sustained focus, because most HSC exams are three hours long. The optimal study plan would probably contain a mix of shorter sessions (preferably a minimum of 30 minutes) and longer sessions, to accommodate smaller and larger tasks, allowing students to build focus and stamina while managing mental fatigue.

Priority:

In all assessment preparation lists there are some items which are essential, others which are beneficial but less crucial, and some that are more time sensitive than others. We talked about prioritising in an earlier chapter, but to refresh your memory – prioritising involves balancing the relative importance of tasks with their urgency. Top priorities are both important and urgent, high priorities are important but less urgent, lower priorities are not important but seem urgent, and the lowest priorities are those activities which are neither important nor urgent. A good plan should always be constructed with priorities in mind, guided by the 80/20 rule, so if you run out of time and can't complete everything on your list you can be confident you have done all the high value essentials, and whatever remains incomplete is not going to cost you significantly. For example, in maths preparation it's crucial to do past papers and revision provided by the teacher, but if you don't have time, it's perhaps less vital to do the easy questions in the textbook chapter reviews. In chemistry, it's essential to make and revise notes and do past papers, but perhaps less important to do exercises from the activity book. In Year 12 English, it's essential to create and revise a master essay and practise adapting, and perhaps less important to memorise the whole quote bank.

When scheduling your preparation, always populate your plan with the **highest priority** items **first** and work through the list allocating the lowest priority tasks last. Ideally, you should aim to incorporate all tasks somewhere in the plan, although there are times that the sheer volume of revision resources makes this objectively impossible. Either way, a consideration of relative priorities ensures that the most important items will get done. Just to be clear, this doesn't mean you place your highest priority items first chronologically, but rather you will allocate these to the most **strategic** timeslots first and then fill in the gaps with lower priority tasks. When scheduling tasks of equally high priority, give preference to those which are most challenging, time consuming, and fill you with the most dread. Getting these out the way early is very helpful.

A note about prioritising **past papers**:

It's wise to think strategically about the timing of your various past papers when planning. This is a high priority category of preparation and within this, the most recent papers will be most important, and the oldest paper will be lower priority. It's ideal to strategically place the most recent past paper close (but not too close) to your exam, as this is the best, most relevant practice ... but you also don't want it to be the last paper in your plan, because you don't want to risk running out of time to do it. We always planned the most recent past paper a minimum of four days before the exam, leaving enough time to guarantee it could be done, while also leaving enough time to resolve any questions. We then planned in the oldest paper afterwards. It would almost always still be done, and constituted valuable final practice, but if it absolutely had to fall off the radar, it wouldn't be a catastrophe.

Planning tasks in priority order aligns with the 80/20 rule because the highest priority preparation will generate most of your results, and the less vital tasks, although still beneficial, contribute proportionally less to the outcome. Having said that, students who are aiming for top marks should try to do everything on the list – and perhaps more.

Additional note regarding planning and prioritising for multiple assessments:

When creating a study plan for a single assessment, the process we have outlined is fairly simple to implement, but when you are planning for more than one assessment and multiple deadlines it gets a little more complicated.

Most of the process remains the same, but instead of prioritising and planning **one** to-do list, you need to **prioritise and plan multiple lists** concurrently, integrating them into one **overall study plan**. This introduces a few extra considerations:

- The relative weightings of the assessment tasks (if there is a differential):
  - For example, if you are planning simultaneously for an English assessment that's weighted 25% and a geography assessment that's weighted 10%, you will give a bit more priority to the English preparation, and this will be reflected in the time allocation and strategic composition of your plan.
- The relative timing of the assessment tasks:
  - For example, if you have six exams spread across three weeks, you should prioritise preparation for the first exams earlier in your study plan, because you will have additional time to prepare for the later exams. This does **not** mean you should neglect the later subjects initially, it just means you need to be conscious of the relative urgency of preparation for different subjects and accommodate for this in your plan.
- The volume of preparation required for different assessments:
  - If there are some assessments with relatively fewer preparation requirements and others with onerous and time-consuming requirements, you need to account for this when prioritising and allocating time.
- Your relative strengths and weaknesses:
  - It's a good idea to prioritise your weaker subject areas to build skill and understanding, as this improvement will

positively impact your overall performance ... BUT don't neglect your stronger areas as these still need significant attention to generate academic advantage.

- The need to prioritise **within** AND **between** subjects:
  - o You might want to conceptualise your individual subject priority lists in terms of one giant master list, using this as the foundation for planning, scheduling the highest overall priority items first and working your way through to the least important items. Your final study plan will be a comprehensive integration of your assessment preparation for all subjects.

A final note about prioritising:

Always return to the basic principles of **time management** which must account for both the **importance and urgency** of any given task. The hope is that most things in your study plan will be Quadrant 2 activities, but the general rule is that if time is running out and some of the most important tasks are becoming urgent and moving into Quadrant 1, these should become your top priority.

A small disclaimer: I have outlined various tasks and priorities here from our experience, but these may not reflect the priorities of all students, depending on many individual factors. The key learning here lies in the application of the **strategic process and considerations**, not the specific examples given.

<u>Evaluating logic:</u>

When you are busy and stressed, logic in the planning phase can sometimes escape you, so it's important to have this at the forefront of your mind when scheduling tasks. These are a few things to check for:

- Ensure the plan follows a logical sequence – for instance, it makes no sense to schedule in a notes revision session before you have made your notes. A logical sequence means first making notes, then revising, then doing past papers and further revision (depending on the nature of the assessment and the subject). There are some tasks which naturally fall either earlier or later in the process, so it's important to factor this into your plan.

- Ensure your plan makes sense – for instance, it might not be sensible to devote all your study time to maths and economics the day before your chemistry exam.
- Ensure your study time is appropriately apportioned to cater to all your subjects – for instance, if you are preparing for three equally weighted subject exams but you have allocated 70% of your study time to one subject, you might need to revisit your overall priorities and revise the plan to reflect a more equal distribution.
- Ensure your plan is balanced throughout, so you can progress your preparation for all subjects – for instance, if preparing for multiple subjects it would be inadvisable to complete six biology past papers before starting on past papers for any other subject.

These points may seem intuitive and be easily dismissed, but when populating plans and working at the micro level of assigning tasks to timeslots, it's easy to lose sight of the big picture. It's often only when the plan is complete (or underway) that lapses in logic reveal themselves. Alternatively, plans can be logical at the start, but if things need to be rescheduled this can compromise the overall logic of the plan. Once you have scheduled in all the tasks on your to-do lists, it's advisable to take a step back and **evaluate** the plan holistically, checking it makes logical sense and adapting if necessary.

TIP

Try to reserve the day before an exam for study specifically related to that subject.

The day before an exam should ideally be devoted to that specific subject, applying single-minded focus to final revision, practice, consolidation, and attending to last-minute preparation details. It can be helpful if you plan those days in first before populating the rest of your plan. This is one of the reasons we found it easier to plan backwards. Note: Although ideal, this may not be possible if you have multiple subject exams on consecutive days.

TIP

Leave a few gaps in your plan if you can.

You should ideally plan gaps into your schedule each day, giving yourself breathing space and defined breaks for exercise and guilt-free relaxation – although this can be difficult to adhere to, especially before big assessment blocks, yearly exams, and HSC trial and final exams. At the very least, you should try to balance your weekly timetable with a couple of 'lighter' days. This helps create a buffer in case you have work overflow from previous days and hopefully allows for a bit of downtime.

TIP

Colour code your plan by subject.

This may sound trivial, but it's a great (and easy) way to assess the balance of your plan at a glance. Assigning a different colour for each subject makes it visually clear whether you are balancing your study appropriately across subjects and makes it easier to evaluate the overall logic.

TIP

A plan is not cast in stone – it needs to be reviewed and revised!

Even the best plans should be re-assessed over time. Things seldom play out **exactly** as intended for a variety of reasons. It's thus vital to track your plan in terms of progress, priorities, and time management, making the necessary adjustments when required.

For example, if a portion of work takes less time than anticipated, you may want to pull another task forward to alleviate pressure later. Conversely, if a task takes longer than expected, or you are unable to complete all scheduled tasks on a certain day, you will need to find time for this on another day, which will necessitate some juggling (while still accounting for overall priorities).

It's also advisable to be aware of emerging 'patterns' and replan to accommodate these. For instance if you have planned four hours to complete each HSC past paper, which includes the 3-hour time limit plus an estimated one hour for marking, and you find that this process is regularly taking you 4.5 hours, then you should adjust the timing in your plan to 4.5 hours for all similar past papers in the future.

You might also find that as you work towards your deadlines, there is a need to reprioritise certain tasks as relative importance and time sensitivity shift.

TIP

Plan according to the 80/20 rule!

While it's important to make and revise your plan, it does not need to be perfect. A 'good' plan that helps you progress with your workload is far better than a 'perfect' plan which takes up so much time that it prevents you from actually getting the work done.

3. Additional planning TIPS for exam blocks, Year 12 Trials and HSC

The assessment planning techniques and tips outlined so far can be applied to any year, but there are extra considerations when planning for big exam blocks with multiple competing deadlines, especially those at the end of Year 11 and throughout Year 12.

TIP

You will likely need to plan study time in the school holidays from the start of Year 12.

If you have assessments early in the term, it's advisable to create a study plan for the preceding school holidays. If you have assessment notifications, you will follow the usual planning process, but in the absence of these it's recommended to utilise some holiday time to ensure you are up to date with work and study notes for all your subjects, while balancing this with planned relaxation breaks.

In terms of timing: the July holiday in your HSC year often precedes trial exams (depending on your school exam schedule), and the September 'holiday' falls immediately before the HSC final exams. When planning for these intense exam periods, it's important to pace yourself.

TIP

Create long-range plans where appropriate.

While most assessments throughout high school involve plans spanning only a few weeks, in Years 11 and 12 you sometimes require plans which span 6–8 weeks (or more), particularly around prelim exams, trials, and final HSC exams.

Longer-range plans are beneficial because you will be studying for most of your exams concurrently, your to-do lists will be significantly longer, and the volume and complexity of content will be substantial. Good organisational skills and a longer lead time will give you the best chance of tackling everything on your to-do lists, creating academic advantage and enhancing your chances of academic success.

TIPS and strategies for building your longer-range Trial and HSC study plans

- Plan to start preparing and revising well in advance of exams – the more revision you do, and the more familiar you are with the content, the better prepared you'll be.
- Space out your study to build long-term memory – e.g. 10 x 1-hr study sessions on a section of content is more beneficial for information retention than 1 x 10-hr session.
- When planning study sessions, try to schedule a mix of subjects each day – this keeps your preparation balanced across subjects, ensuring you are progressing in all areas and not surging ahead in one subject while neglecting others – except the day before an exam, when you would ideally focus on the relevant subject.
- Try to schedule a mix of task types and a balance of your STEM and humanities subjects each day, as this engages different parts of your brain, enhances focus, and reduces mental fatigue – for example, instead of planning three maths papers in a day, rather plan for one math paper, an English essay and some economics revision.
- Try to schedule a mix of shorter and longer study sessions, populated with appropriately timed tasks.
- English should always be high priority in your study plan as it's compulsory for HSC and will definitely contribute to your ATAR. Try to schedule English in every day (or at least every second day) in the period before exams. Although setting up your master essays can take time, hopefully this would already have been done before the lead-up to exams, and revision and practice sessions can be reasonably short, as a practice essay only takes about 40 minutes and revision doesn't take long.
- If you aren't sure which of the day's tasks to tackle first, start with the highest priority ones, and within that, the most difficult, as it feels good to get them done.
- Plan to start doing past papers as soon as reasonably possible because most trial and HSC past papers are three hours

long. If you have 5–6 subjects and you intend to do **at least** six past HSC papers (or equivalent) for each subject, plus marking, that adds up to a lot of hours. If you plan to do them in exam conditions (which is advisable), and you haven't started early and planned well, you will run out of time to get them all done.

- When scheduling past papers into your plan, try to mirror your trial/HSC timetable as far as possible – e.g. if you have your maths exam at 2pm, try to plan some of your past papers at this same time. It's not ideal doing all your maths practice in the mornings when you are fresh and then sitting the final exam in the afternoon when you are usually more tired, having not practised under these conditions. The more you replicate exam conditions, the better your training will be.
- If possible, factor in time to request feedback from teachers on your practice essays, responses, and past papers. If your teacher is willing to do this, it requires extra lead time in your plan so they have time to mark, and you have time to apply feedback.
- Aim to do a certain number of hours of study per day but allow some flexibility and try not to be unrealistic.
- Within the exam period: On exam days, make your study schedule a bit lighter to accommodate for fatigue and reduce stress.
- Try and create a plan that facilitates a routine in terms of start times, mealtimes, finish times, and bedtimes. Routine can be a very helpful tool in managing stress.
- Leave time in your plan for exercise, a bit of socialising, and relaxation.
- Aim for consistency in your plan, adopting a steady and **sustainable** approach.

Note: In some ways, trial exams can be more stressful and difficult to plan and prepare for than the HSC finals, because during trials you are still juggling school, homework, and all your usual commitments. It's harder to find the space to study, and many teachers are rushing to complete content that will be examined. It's a very intense and overwhelming time, which makes planning even more crucial if you are to optimise your available time and resources. After trials, things ease up for a bit, and everything you have prepared for in trials will be ready to use again in your HSC exams. But this is not a time to become complacent ...

<u>TIP</u>

Plan to start your HSC study preparation as soon as reasonably possible after trials.

Although some schools hold their trial exams earlier, most schedule these soon after the July school holiday. Once trials are finished, there are usually approximately 6–8 weeks until final HSC exams. It can be very tempting for students to take their foot off the accelerator after this for a variety of reasons (some of which are very valid):

- Most of the Year 12 content has (hopefully) already been learnt and prepared
- Students are exhausted after trials (because these exams are gruelling)
- Final exams seem far away
- There are graduation and end of Year 12 celebrations

It's not an issue for students to take a few days off to catch their breath and enjoy the festivities, but too much of a break can be problematic because:

- Teachers are covering final sections of content and there is still work to be done.
- The last few weeks of school are very busy, disruptive, and go very fast.

- It's hard to get back into study after a break – the longer the break, the harder it is.
- If too much time passes your memory fades, you lose the benefit of recency, and you will have to relearn content.
- It's easy to underestimate how long the final preparation will take, and once school finishes and graduation is complete, you only have 2–3 weeks left to do everything.

The best advice is to take the weekend off after trials to rest and recover, then create a plan for the remainder of term, the study break, and the exam period (factoring in the end of term and graduation events).

Here are two real examples of HSC workplans, incorporating many of the tips above:

Day	22-Sep	23-Sep	24-Sep	25-Sep	26-Sep	27-Sep	28-Sep
8-9						English essay Module B	
9-10	Physics 2020 HSC	Mark French listening	Chemistry 2021 HSC	Math X2 2021 HSC	French 2020 HSC	Math X1 2020 HSC	Physics 2021 HSC
10-11	Physics 2020 HSC	French writing practice	Chemistry 2021 HSC	Math X2 2021 HSC	French 2020 HSC	Math X1 2020 HSC	Physics 2021 HSC
11-12	Physics 2020 HSC	French writing practice	Chemistry 2021 HSC	Math X2 2021 HSC	French 2020 HSC	Mark Math X1 2020 HSC	Physics 2021 HSC
12-1	Mark Physics 2020	Maths X1 hard textbook Q's	Mark Chemistry 2021	Mark math X2 2021	Mark French 2020	Math X1 past HSC q's	Mark Physics 2021
1-2	lunchbreak	lunchbreak	lunchbreak	lunchbreak	lunchbreak	lunchbreak	lunchbreak
2-3	French listening	English review class notes	Physics notes revision	Physics revision	Chemistry notes revision	Physics practical revsion	Math X2 Past HSC q's
3-4		English essay Module A	Physics notes revision		Chemistry notes revision	Physics notes revision	Math X2 Past HSC q's
4-5	English practice comprehension			Chemistry practical revision			
5-6	Math X1 extra teacher revision	Mark French writing	English master essays revision		English master essays revision	English master essays revision	English master essays revision
6-7							

Source: Study plan courtesy of A.A. (ATAR 99.5)

Figure 9.1 Example HSC study plan

Monday 29/09	Tuesday 30/09	Wednesday 01/10	Thursday 02/10	Friday 03/10	Saturday 04/10	Sunday 05/10
☐ Revise English Module C creative (10 min) ☐ Practice Mod C creative (40 min) ☐ Chemistry 2020 HSC paper + mark (4 hrs) ☐ Math Ex1 2022 HSC paper + mark (3 hrs) ☐ Math tutor 4:00-6:00	☐ Modern history school tutorial (2 hrs) ☐ Biology 2021 HSC paper + mark (4 hrs) ☐ Finalise mod A English paragraphs (1 hr) ☐ English revise Module A (20 min) ☐ Practice essay Mod A (40 min)	☐ Revise English Module B essay (10 min) ☐ Practice English essay Mod B (40 min) ☐ Chemistry 2022 HSC paper + mark (4 hrs) ☐ FINISH Modern History essay SCAFFOLDS (1.5 hrs) ☐ Biology revision module 7 again (1.5 hrs)	☐ Math Extension 2 2022 HSC paper + mark (4 hrs) ☐ Modern history practice essays Indochine (1.5 hrs) ☐ Revise all english (1 hr)	☐ 9:00-2:00 English full mock paper 1 & paper 2 (5 hrs + travel time) ☐ Revise english (1 hr) ☐ Modern History Revise all civil rights (2 hrs)	☐ Revise English (1 hr) ☐ English practice essay Mod A (40 min) ☐ Math Extension 1 2023 HSC paper + mark (3 hrs) ☐ Modern History Power and Authority practice (1 hr) ☐ Revise again chemistry module 5 (1 hr)	☐ Biology 2022 HSC paper + mark (4 hrs) ☐ Math Extension 2 2023 HSC paper + mark (4 hrs) ☐ USSR practice essay (45 min)

Source: Study plan courtesy of I.A. (ATAR 99.8)

Figure 9.2 Example HSC study plan

Things to note in both plans:

- Both plans have been constructed using the **'S.T.E.P.'** principles (**Specificity, Time management, Evaluating logic,** and **Priority**).
- Colour-coding subjects helps to **Evaluate logical** flow and balance at a glance.
- Tasks and their estimated timings are **Specific**.
- Tasks are **Prioritised**.
- **Time management** is clear and achievable, with the odd shorter day creating a time buffer to take a break or accommodate incomplete or additional work.
- Past papers are interspersed with additional revision.
- English is represented almost every day as a priority.
- Each day contains a mix of subjects, task lengths, and task types.
- Allowance is made to build stamina and sustained focus for longer exams.

A few caveats:

- These study plans each only show a single week after graduation (during the September study break), approximately two weeks prior to the start of HSC exams, so any perceived imbalances would be accommodated for in prior or subsequent weeks. A proper evaluation of the plan would therefore need to consider the entire schedule rather than this small excerpt.
- In this particular time period, no other activities (except for tutoring) have been included, because by this stage most other commitments had concluded and all available time could be devoted to study.
- While these are excerpts of real plans, they are specific to the individuals and are included solely for illustration purposes. They would **not** be directly applicable to other students, as study plans need to be tailored according to a student's specific subjects, available time, personal commitments, priorities, study methods, goals, resources, and preferences.

TIP

Stick to your plan, remain focused, employ self-discipline … and avoid procrastination!

Try to stick to your plan but if you can see early on that it's not feasible, then replan to a more sustainable version. You are more likely to stay on track if you remain focused during dedicated study times and avoid procrastination as much as possible. Self-discipline is key! Procrastination and work avoidance is a very common problem, especially when students feel overwhelmed during periods of stress and pressure, but it can land you in a destructive cycle. Falling behind your work plan creates additional stress and can prompt further procrastination putting you in an unhelpful spiral.

To minimise the temptation for procrastination, keep your phone and other distractions away from your study area and try to take your breaks in a different space.

It's also helpful to spend a few minutes at the end of each day organising your resources for the following day. This helps you to start your next day according to plan, minimising disruption to your schedule and keeping you on track.

TIP

Study apps can be helpful (sometimes).

Using a study app can have benefits if used constructively:

- Study apps which track progress and provide statistics showing how much time you are spending on each subject can be very helpful. This allows you to strategically evaluate and refine your study plan, ensuring you are allocating your time appropriately with a holistic view.

– Study apps which connect you with peers can be beneficial, creating a sense of community. Studying for the HSC can be immensely stressful and isolating for many students. Connecting with other HSC students 'in the same boat' can be helpful. Knowing that others are going through the same experience, and seeing when peers are also studying can reduce the sense of isolation and improve motivation.

***

In this chapter we have covered the important role of planning in academic success. I hope these tips, strategies, and the step-by-step guide for creating effective assessment study plans will assist you in optimising your resources to secure academic advantage. If you work smart and employ these planning techniques, maximising the time in Quadrant 2, you should be well placed to successfully manage the intense workload.

***

We have almost reached the end, having discussed everything from starting Year 7 to planning for the final HSC exam. There is a tremendous amount to digest and, as I said early on, everything contained in these pages is based on things we have learnt, both personally and vicariously, that have been instrumental in generating academic success at the highest levels. In this next chapter, I thought it would be helpful to take a brief look at some of the strategic decisions you might need to make through high school, and how you might approach these in a way that gives you an academic edge.

CHAPTER 10:
Making strategic choices when the answer is not clear-cut

Students are confronted with various choices through high school, many of which have a clear-cut 'correct' option. For instance, I don't think there would be any debate about whether students should pay attention in class, study diligently, learn how to answer the questions, or revise before an exam. The right choice in such examples is usually obvious and unambiguous. There are, however, certain scenarios which are less clear-cut, having no definite right or wrong, and the final decision of which path to take may rest on an assessment of relative pros and cons, with the optimal strategy varying by student. Since most students will have to make a few choices within these grey areas, it's worth taking a moment to consider the strategic implications of various options.

With that said, let's discuss a few key decisions that students will face, along with some of the main considerations:

1. Subject choices and electives

After broad exposure in Year 7, the first significant subject choices are made during Year 8 for Year 9. English, maths, science, PDHPE, history, and geography currently remain mandatory for Years 9–10, and students choose additional elective courses tailored to their strengths and interests, while dropping some of the subjects they don't connect with. Examples of electives might include languages, commerce, art, drama, music, technology, elective history/geography courses, among a host of others. This provides an opportunity to try subjects before making decisions about HSC. The caveat is that not all subjects are offered at all schools, as this depends on the size of the school and classes, demand for subjects, timetable issues, or specialist teacher availability.

TIP

Before choosing Year 9 electives, consider what you may want to study in Years 11–12.

When choosing Year 9 electives it's important to consider the potential impact on Year 11 and 12 study options. Years 9–10 electives are usually chosen in the middle of Year 8, and the HSC seems far away. However, choices made for Year 9 may have repercussions for HSC subjects. For example, subjects such as language continuers and music are difficult to pick up in Year 11 and 12 if they were not chosen in Years 9–10. Eligibility would usually have to be discussed with the school on a case-by-case basis, but in any event, it would likely create disadvantage, especially if the rest of the subject cohort has two years of prior learning and practice. Therefore, if you would like the option to choose a subject for HSC, it's important to find out if you should choose it for Year 9.

There are other subjects which don't have the same continuity implications, meaning if you don't choose them for Years 9–10, you can still pick them up later with no issues. That said, there may be some advantage for students who have done an additional two years of their chosen subject, relative to students picking it up for the first time in Year 11. For example: the Years 9–10 commerce course has various modules including economics, business, and legal. It's beneficial for students to select commerce in Years 9–10 if they are considering taking economics, business studies, or legal studies for HSC, because the taster modules provide a great opportunity to 'try them out', and if they choose to pursue these subjects later, they already have a foundational understanding. However, it's entirely possible to pick these subjects up later without having chosen commerce as a Year 9 elective – it might just require a little catch-up.

> **TIP**
> **Choose Year 11 subjects with your Year 12 strategy in mind.**

The next fork in the road comes in the middle of Year 10 when students pick the subjects they will study in Year 11. In Year 11, students are required to study at least 12 units, which usually equates to six subjects, or perhaps more, depending on the mix of courses. Additional specialised subjects such as economics, business and legal studies, engineering, and many others, also become available. Science splits into the specific fields of biology, chemistry, physics, earth and environmental science, and certain extension courses become available for maths and English. The only subject that is compulsory throughout high school and for the HSC is English.

Subjects chosen for Year 11 should be carefully considered as they ultimately determine your Year 12 options. The only 'new' subjects you can generally pick up in Year 12 are additional extension courses for subjects you are already doing (if you meet the criteria). So, before committing to Year 11 subjects, you should be clear on the subjects you want to study in Year 12. For example, if you have not chosen to do at least one of the sciences in Year 11, you won't be able to choose Science Extension in Year 12. Similarly, if you haven't done Maths Extension 1 in Year 11, you won't be eligible to do Maths Extension 1 or Extension 2 in Year 12. So, ensure that the subjects chosen in Year 11 allow for your preferred subject mix in Year 12. If in doubt or you need guidance, it's advisable to discuss with the school before making your final choices.

TIP

Choose Year 12 courses with consideration for Year 11 performance and enjoyment.

By the time you make final Year 12 subject selections, around the middle of Year 11, you should have a good idea which direction to take based on your enjoyment and performance in your various courses. For Year 12, further extension options become available for English, maths, science, history, languages, and a few other subjects. While you can't completely change direction for Year 12, you can tweak and finetune your subject mix to align with your strengths, by dropping some courses and picking up others. For example, if you didn't enjoy chemistry but you thoroughly enjoyed French and history, then you might choose to drop chemistry and pick up additional extension units in history and French. Or if you did not perform well in Maths Extension 1, but you excelled at English Extension 1, you might opt to drop Maths Extension 1 and pick up English Extension 2. Your final Year 12 subject choices should be based on what you enjoyed and did well in throughout Year 11.

TIP

Don't make subject choices based solely on their perceived scaling potential.

By the time you make Year 11 and 12 subject choices, you will have heard a lot of chatter about choosing subjects that 'scale well' in the HSC. Schools almost always advise students not to worry about scaling, and to rather choose subjects they are good at and interested in. This is absolutely correct!

While it's true that certain subjects tend to scale better than others (usually the more challenging ones), it's not advisable to choose a subject just because it may scale well, **unless** you are confident in your ability to perform well. Excellent performance in a 'lower' scaling subject will likely serve you better than poor performance in a 'high' scaling subject. My daughter is a good example of the positive impact of excelling in a 'lower' scaling subject – alongside her mix of challenging STEM courses, she also chose modern history, which was comparatively lower scaling. The choice was motivated by enjoyment of the subject, and despite lower scaling, she went on to state rank, meaning modern history contributed positively towards her ATAR, on equal footing with her Band 6s in all her high-scaling subjects.

Your best strategy should therefore be to select subjects which you enjoy and are more likely to excel in. This is far more important and beneficial to your end results than choosing subjects based purely on what you believe about scaling. It's also important to remember that scaling of subjects can change from year to year depending on many factors, so using this as your only selection criterion is a flawed approach.

The only time that scaling should factor into your decision-making process is if you are equally confident and capable in each of the subjects you are weighing up – then it's advantageous to choose the one that might scale better.

TIP

Being strategic with subject choices does not mean taking the easy way out.

In light of the previous tip, students might debate whether it's better to choose an easier course they can **excel** in, or a harder course they can **do well** in. This is an individual decision and there may be many personal factors which need to be considered, but the general advice is to take the course that **suits your ability level** and provides a healthy level of challenge, rather than a course that's **too easy** for you.

For example, if you are a capable maths student and you can perform well in advanced maths, it's better **not** to opt for standard maths, even if your absolute mark may be higher. In the situation where you are capable of performing well in both courses, scaling can be beneficial, because you need a higher mark in standard maths than advanced maths for the same ATAR contribution.

HSC scaling is designed so students choosing harder courses are not disadvantaged by doing so. The system therefore encourages students to choose courses **suitable for their ability**, and if these happen to be harder, scaling accounts for this.

2. Deciding whether to do 10 or more units for HSC

Each HSC course is allocated a certain number of units. Most subject courses are worth 2 units, but there are some which are worth 1 unit. Students can mix and match a combination of 2 unit and 1 unit courses (within NESA guidelines), but all students completing their HSC are required to do a minimum of 10 units, of which 2 must be an approved English course.

Once the 10 unit minimum has been met, students have the option to do additional units. Only a minority of students will opt to do as many as 13 units, which means most students will be deciding whether to do 10, 11, or 12 units.

It's important to note that although you receive HSC results for all your units of study, the ATAR is calculated from your 10 best contributing units (2 of which must be English), and any additional units will not contribute to the ATAR.

Some schools encourage students to do 10 units, while others encourage 12 units. If the student has an option, there are a few pros and cons to consider.

The **pros of doing 10 units** in Year 12 (which are also cons of doing extra units):

- You can focus all your attention on the 10 units that will count towards your ATAR
- You have more time to allocate to each subject if you aren't doing extra units
- You don't waste time on units that won't end up counting
- You give yourself the best chance to excel in fewer units
- You have less classes to attend **and** more free periods in your school timetable
- You have less homework, less study, and less exams
- Overall, you have the dual benefit of less cognitive and academic load **and** more available time.

After hearing all the compelling benefits of doing the minimum number of units, you may wonder why on earth anyone would choose to do extra units … but there are equally compelling reasons to do so …

The **pros of doing extra units** in Year 12 (which are the cons of doing just 10 units):

- You can't predict how your internal school performance will translate in the external exams, and you can't be sure how your exams will go on the day – anyone can have a bad exam! Since the ATAR takes the 10 **best** units, additional units give you a competitive edge, because if anything goes wrong in one subject, you have an extra subject to buffer you (except for 2 units of English which must count).
- You can't predict the scaling effect when making final subject selections, and sometimes a subject you think will count, does not end up counting – in this case you will be thankful you persevered with another that contributed more. Without this, your ATAR would have been lower. I will digress briefly to illustrate this point – my son was doing 12 units, and within

that he was ranked first in French in his school, while his internal chemistry rank was a bit lower. Chemistry was his least favourite subject and the one he felt relatively weaker in. We had many discussions throughout Year 12 regarding the possibility of dropping chemistry, but he ultimately decided to keep it in the mix. On receiving his HSC results, we were shocked to learn that it was French that didn't count and, for unanticipated reasons, his chemistry became a higher contributor, knocking French (which he had 'duxed') out of the ATAR calculation. Had he dropped chemistry and only done 10 units, his ATAR would have been compromised.

- If you are a high achiever, and willing to put in the extra work, doing 12 units gives you a better chance of making it onto the HSC 'All-round Achievers' list. This celebrates students who achieve the top band (Band 6 = over 90%) in 10 or more units of study (regardless of the subjects). Doing extra units gives you wiggle room and a better chance of achieving this, especially for students who struggle more with English. English does **not** have to be in the 10 best units for this particular accolade.
- And finally, the time 'saved' when students choose to reduce to 10 units does not always get channelled back into their remaining 10 units. If this excess capacity is not effectively utilised, they lose the benefit of a reduced workload, squandering the advantages of both options.

The decision of how many units you study in the HSC will be an individual one, guided by personal circumstances, academic aspirations, preferences, and the views of the school. There is no right or wrong decision, but it's advisable to carefully consider and weigh up the pros and cons of all your options when making your final choice. If you aren't sure what is best, you also have the option of starting Year 12 with 12 units and if you find it's not working well, you may be able to drop a unit or two later. Although you can usually drop an extra subject at any time prior to the final HSC exams, it's best not to leave it too late, because if you drop it near the end of the year, you will have done most of the work and lost all the time-saving benefits derived from studying less units.

3. Subject acceleration

Subject acceleration is a bit niche in that it only applies at certain schools and for select subjects and students. However, I believe it's worth mentioning because if faced with the choice, it's good to understand the strategic implications.

Some schools provide the option for select students to accelerate their studies in particular courses, jumping a year (or more) ahead of their cohort, and writing the HSC exam for that subject in advance. Acceleration is usually by invitation, and students are identified based on their performance in the subject.

Acceleration opportunities arise in various subjects such as Latin, studies of religion, and some of the sciences, but by far the most common is maths. Although many of the same acceleration principles apply regardless of the subject, I will focus specifically on the Mathematics Advanced (2 unit) course because both my kids were maths accelerants, so we have personally experienced the upsides and downsides.

Maths acceleration (if offered) is handled differently depending on the school. Some schools identify candidates as early as Year 7, beginning the official acceleration programme in Year 8, while other schools may start their formal programme later in Year 10. The programme structure also varies between schools, with some creating a custom schedule for the curriculum, with accelerants sitting unique exams which aren't aligned with either their year or the year above. Other schools run the programme in parallel, meaning students may study the content for their year and the year above concurrently, sitting exams for both year groups each year. Regardless of the structure, there are some universal pros and cons to participating in the accelerated program. Let's look at a few of these in more detail ...

Pros of acceleration:

- Students in the class usually have similarly high ability, enabling the teacher to cover content at a faster pace.
- Sitting the HSC early for one subject is a great experience, building confidence and providing practice which can benefit you in your broader HSC the following year.
- If you do well in the accelerated HSC exam, you can use this mark the following year, securing an early boost to your ATAR.
- If you do well and choose to count the accelerated course as one of your subjects in Year 12, it creates additional options for study patterns, as you may be able to take less units in your actual HSC year (since a couple of units will already be done and banked). This can reduce your workload in Year 12, which is beneficial, allowing study time to be diverted to other subjects.
- If you don't perform well in the early HSC, you have lost nothing and can still have a second go, retaking the HSC exam with your original cohort the following year.
- If you choose to do Extension 2 Maths (4 unit maths) in your own HSC year, your accelerated Maths Advanced HSC result will unfortunately not count towards your units or ATAR and will fall away (which can be a con), but having already completed the Advanced coursework reduces the amount of foundational content you need to cover in the Extension 2 course (which is a pro).
- The class usually contains the same students each year, which can be a bonding experience (but could also have a downside if there are difficult class dynamics).
- Having accelerated, especially if you have done well, looks great on your resume.

Cons of acceleration:

- A faster pace with challenging content can create additional stress and pressure.
- As greater volume and complexity of content is covered in each lesson, the homework is usually more difficult and time consuming.
- An accelerated course increases your overall workload.
- You may need to do extra exams and assessments.
- Accelerants sometimes fall through the cracks between year groups, and communication can be patchy – which can add stress and confusion.
- With both the unique and concurrent maths programmes, content can sometimes slip off the radar and gaps in learning may occasionally occur.
- When learning happens far in advance, you may lose some of the recency benefit, making it harder to recall assumed knowledge for later extension courses.
- Exam timetables can be tough to navigate because your exams straddle the timetable for your own year and the year ahead – e.g. you may sit your Year 12 trial exam for accelerated maths while simultaneously preparing for your Year 11 finals for all other subjects. Effective and disciplined time management will thus be critical.
- You will most likely get less holiday because while your Year 10 cohort is enjoying their long summer holiday, you will be studying for your maths exam in Term 1 (which is Term 2 for Year 12), and while your Year 11 cohort is enjoying the September break before starting Year 12, you will be doing your final preparation for your Mathematics Advanced HSC final exam which takes place just after the holiday.
- There is a risk that doing too much too soon can induce burnout if not managed.

When weighing up whether to accept an invitation to join an accelerated class, there is a lot to consider. Once again there is no right or wrong answer, and the final decision will depend on your personal circumstances, priorities, and academic goals. I can say that although the process was challenging and frustrating at times, it was a worthwhile experience with more benefits than costs, and if we had to go back in time, I believe both my kids would still opt to accelerate.

4. Deciding whether to get a tutor – or not

One of the most controversial topics you will encounter through high school is whether to get academic tutoring … and the decision is not always straightforward.

I initially (incorrectly) assumed that the main role of tutors was to provide learning support for kids that were struggling academically … but when my kids were still in primary school it became apparent that many high-achieving students were engaging the help of tutors to assist them in performing **even better**. At the time I was surprised to learn that a significant number of children already had tutors from as early as kindy.

In full disclosure, I thought this was ridiculous, and defiantly declared that neither of my kids would ever be tutored. I genuinely believed that bright kids would thrive on their own, that hard work would be enough, and I saw no reason to subject them to hours of coaching and the additional stress and pressure that came with it (not to mention the expense). I held out for a long time, resisting the push towards tutoring. Our kids were doing well academically, despite being almost the only ones in their respective cohorts not receiving tutoring, but we (them and me) eventually started feeling the pressure, and in the end … we caved! Finally, when my son was in Year 10, we got him an English tutor, followed by a maths tutor in Year 11, and soon after that he enrolled in tutoring for chemistry, and eventually physics. By the time he started Year 12, the only subject he was not tutored in was French. Once we had started on this trajectory and seen the benefits, we decided to get an earlier start with my daughter, who began with the English tutor at the end of Year 8 (prompted by her 'Shakespearean tragedy' experience, which you might recall from my *Romeo and Juliet* assessment anecdote in Chapter 7). She started with a maths tutor at the end of Year 9 and enrolled in chemistry tuition for Years 11 and 12.

Having been vehemently opposed to tutoring for high-achieving kids, and now having experienced both sides, I can say with confidence that both my kids benefited from having tutors. There are, however, some pros and cons which warrant consideration …

Pros of tutoring:

- You cover the content twice (once at school and once with the tutor) which creates greater familiarity and depth of understanding, with less potential learning gaps because between school and tutoring you will learn it somewhere.
- If you miss a class at school (e.g. if you are sick or on an excursion), you still have the safety net of covering it with the tutor.
- If you have a weak schoolteacher (which can happen), you have security knowing there is another source of learning.
- Individual and small group tutors teach faster with more individual attention.
- You are exposed to a greater variety of techniques, question types, and strategies, which expand your toolkit.
- You have access to a broader range of resources and practice materials.
- You are forced to spend extra hours each week working on the relevant subjects, which may not otherwise be done – this develops skills, promotes revision, and practice, all building towards your 10,000 hours.
- You can ask questions or clarify issues you may not have time or opportunity to raise in class.

- Tutors can provide extension opportunities beyond what might be offered in class.
- Lessons can be tailored to the specific needs of the student.
- Students gain access to additional personalised guidance and feedback, which might not be available at school.
- Tutoring can help you to develop strategy, perform better and access higher marks.
- Tutoring provides you with an academic advantage, but equally, NOT having tutoring can sometimes put you at a disadvantage.
- It helps level the playing field if your cohort is largely tutored – on this point, I believe that if nobody was tutored it would be fine, but if most high-achieving students are tutored, it raises the bar for everyone, and those without a tutor can be significantly disadvantaged. In fact, I would say for students aspiring to achieve top results, tutoring is a necessity. (There may be a few students who can dominate academically without tutor assistance but these are, in my experience, exceptions.)
- Some tutors offer online options which is very convenient.
- Some tutors can help holistically with study planning.
- Tutors can be a source of support, encouragement, and confidence building.

Cons of tutoring:

- It can be expensive.
- Some schools discourage it.
- It can be hard to find tutors as some students and parents gatekeep information.
- The high demand for tutors means it may be difficult to get a spot, and you might end up having to do lessons at 11pm or getting stuck on a long waitlist.
- Tutor quality and skill level can be variable – some are excellent and others not – and sometimes it can be hard to know the difference.
- It takes up a lot of time with lessons anywhere from 1–3 hours in duration (plus potential travel time and tutoring homework).

- It can put students under more pressure.
- If students aren't willing to work hard, the tutoring benefit will be marginal.
- There is a risk that tutored students might zone out at school on the assumption they already know the material, or they may neglect schoolwork in favour of tutoring homework – but this is student dependent.
- If tutors teach too far ahead of the curriculum, it can be confusing and counterproductive.
- If students don't buy into the process and are forced into tutoring it can cause conflict and diminish learning opportunities.

Having illuminated some pros and cons, there are a few other considerations when assessing whether to engage a tutor or deciding what type of tutoring model to pursue:

Individual vs class tuition:

There are benefits to both models, and it largely depends on the subject in question and the needs of the student.

- **Classes** run by formal tuition centres (which often contain a mix of different schools) tend to be better at running structured programmes with structured resources, building general subject skills, drilling techniques, and running mock exams. However, in catering for a mix of schools, the lesson plans and specific aspects of content may not always coincide optimally with the schedule of teaching or exams at each school. Furthermore, if classes are too big, some of the tuition benefits are compromised. The ideal class size would be ten or less students. In class tuition, students are not usually able to customise the programme to suit their specific needs, and individual guidance and feedback may not be as readily available.
- **Individual** tuition is better for flexible programmes which are tailored to the specific needs of the student, providing personalised

feedback and guidance. The teaching schedule can be aligned with the school programme, and the pace and focus of lessons is adaptable to suit the student. Flexible, personalised tutoring is especially beneficial when the student is working on content specific to their school, such as an English set work, or if a student needs help with their assessment preparation. On the downside, individual tuition can sometimes lack the structure, resources, exam practice, and technique building opportunities provided in classes.

Professional tutor vs casual tutor/university student:

There is a place for both professional/career tutors and less formal tuition provided by university students and casual tutors.

- **Professional** tutors usually have more experience teaching and communicating, and a greater level of subject expertise. They would also usually have access to a greater breadth of resources. Given the demand for experienced tutors, it can be difficult to secure a spot, and they can be more expensive.
- **Casual** tutors, often university students, may not have the same teaching experience, but are usually happy to share their school results, which can provide some assurance about their grasp of the subject and overall academic capability. Uni students, being generally younger, have the added advantages of relatability and recency regarding subject knowledge and familiarity with the curriculum. They also tend to be more flexible with lesson times, and more moderately priced.

Does the subject warrant tuition?

Not all subjects require tuition, unless of course the student is struggling.

The subjects deriving the greatest benefits from tutoring are usually those which are:

– Complex and challenging
– Abstract or conceptually driven
– Reliant on specific skill development and exam techniques
– Focused on problem-solving
– Application based
– Heavily analytical

The subjects with the highest demand for tutoring tend to be maths (all levels), English (most levels), physics, chemistry, and economics. In our experience, subjects such as history, biology, and an array of others, do not generally require extra tuition because although they might be content heavy, they are not particularly complex. Having said that, many tuition companies offer classes for a variety of subjects, so if a student is struggling, the help is available.

In the end, the tutoring decision depends on the individual circumstances and academic needs of the student.

***

In this section, we have focused on some of the choices students will inevitably face throughout their high school journey. These are not clear-cut and each decision will be made based on the student's personal circumstances and preferences, subject inclinations, priorities, and academic aspirations. The aim of this chapter, which I hope I have achieved, is to highlight some of the decision points and their respective considerations, so that students are informed and well placed to make the most strategically appropriate decision at every turn to maximise their academic potential and achieve their best results.

***

At this stage I think we have covered everything that I initially set out to share (and quite a few additional things that came to mind along the way). But before finishing, I thought it would be valuable to pan out and share some final tips and strategic insights from others to lend a broader perspective …

PART 4:
WRAPPING IT ALL UP

"Self-discipline is what sits you at the desk every day and tells you to keep pushing" (I.A. – 99.8 ATAR.)

CHAPTER 11:
Key insights from top HSC achievers ... and final thoughts

The most valuable advice usually comes from those who have personally endured struggle and triumphed. So, I decided to survey a selection of top HSC performers to get their strategic insights and tips for achieving academic goals, as well as their thoughts on the most valuable and necessary skills. Although these have been written with the HSC in mind, most of the advice and strategies are broadly applicable to all stages of high school.

Although the names of the students will not be disclosed for privacy reasons, they have all authorised me to share their responses and ATAR information. These include a mix of genders and graduating classes as well as a variety of schools. With that said, let's hear what they had to say...

1. Insights from some top HSC achievers

A.A. – ATAR 99.5

Biggest tips for HSC and exams in general:

1. Make sure you begin **revising** content well **in advance** of exams. I know this is said a lot, but a week of extra preparation can push your knowledge to new levels and help you access higher marks for complex exam questions. It improves the detail, accuracy, and speed with which you can answer questions.
2. **Practice, practice, practice**! Having the knowledge is one thing but putting it into practice is another aspect of exam preparation that is sometimes overlooked. Practising how to answer different questions, responding to different directive verbs and even how to plan responses under **exam conditions** can be critical to accessing higher marks in an unseen exam.
3. Maintain a proper **balance**. Some people may view ideal exam preparation as spending all day inside studying, but this is often not the case. Ensuring you still find time to spend with friends, family, and pursuing personal hobbies is imperative to maintaining a healthy mindset in stressful times. It will also keep you refreshed and working at your best for longer.
4. High school is a **marathon, not a sprint**. If you make a couple of mistakes in an exam and do worse than you would have liked, it's not the end of the world. There is always an opportunity to improve and recover!

I.A. – ATAR 99.8

TIPS and strategies:

1. First of all, and probably most important is **planning**. I had many peers who would pull all-nighters before exams, do past papers too late and they were extremely stressed the morning of an exam because they hadn't finished revising or memorising content. Not only did this cause stress, but this disorganised method of studying, while it might yield good results in Years 10 and 11, was the pitfall of many students I knew in their final year of high school. If you make a plan, even two – three weeks before the exam block, and schedule in the past papers and revision that you need to do, it is much less stressful, and personally I think it is the most important way to achieve results.
2. Secondly, **practice!** Especially for subjects that require application of content like maths and even for subjects like English and history where you need to adapt your content to ATQ, by grinding through **practice questions** and **past papers** you are prepared for the type of content you can expect in the exam, and you learn approaches to certain exam-style questions that you may have not known before. For something like maths, revising your notes can only get you so far, and it is useless if you can't actualise your notes into solving a question!
3. Alongside practice and planning, **consistency** in studying is very important. You need to, especially in the lead-up to HSC exams, build up a schedule that keeps you consistent and doesn't give time for you to 'not feel like studying'. For example, in the 'holidays' before HSC, I started studying at 8:00 on the dot every morning, and if I woke up earlier, I would start at 7:30 to provide **routine** and restrict myself from feeling like studying was a choice.
4. I also highly recommend the use of a **study app**. Personally, I used YPT, and I think it was really valuable to have statistics about my studying. I know some people don't like study apps but for me,

it allowed me to push myself harder to achieve specific times of study per day and also track how much time I was wasting during meal breaks. I also found it reassuring and motivating to see other students in my age group studying alongside me, with around 1,000–2,000 HSC students on at the same time as me per day. It is also good if you are a bit competitive like me!

5. The last point is also really important: **seeking feedback and taking on feedback**. Teachers (most of the time) know what HSC markers are looking for, and some teachers are even HSC markers themselves. It is so valuable to be able to improve to the standards of the HSC, but you can only do this when you proactively ask for feedback and apply it. For example, I would make a note document of feedback or mistakes I made when writing responses for subjects like chemistry and biology, and revise this to make sure that before the exam I was applying the feedback of my teachers.

Key skills:

I believe skills like time management, self-discipline and work ethic were the most important contributors to success, especially in the higher grades of high school.

1. **Time management** is extremely important for success. Almost everyone is prone to procrastination but if you manage your time correctly, you can have time to relax at the end of the day after studying. This allowed me to complete all my tasks for the day and still be able to recharge before another day of studying.
2. **Self-discipline** is also extremely important to academic success. I can almost guarantee that no Year 12 student woke up every day for over a year being excited to study. By the end of high school, you lose motivation and become so entrenched in rigorous study that it is hard to stay motivated by your goal ATAR or a greater desire in life. Ultimately, self-discipline is what sits you at the

desk every day for hours on end and tells you to keep pushing. It is crucial for success.

3. Lastly, **work ethic**. This lies along the same lines as self-discipline but, the way I see it, work ethic is also about the sacrifices you have to make if you really truly want that top ATAR. You have to have a really strong, grounded sense of your own study that can't be infiltrated by other people. For me, it was knowing that if I sat down to study with some of my friends, it would be an unproductive waste of time. It also meant consistently prioritising study over social arrangements that would provide instant gratification but cause me longer-term stress because I would fall behind. I know a lot of my peers who deluded themselves, saying they were 'studying' as a group in the library, while I know that the majority of that time was spent chatting, or even just not being able to focus. I strongly advise to either study alone, which minimises distractions, or study with someone equally driven and determined as you, someone who understands that studying is not a social activity.

S.R. – ATAR 99.85

TIPS:

1. **Eliminate** any **distractions**, e.g. phone or YouTube.
2. Attempt as many **practice exams** and **questions** as possible under **timed** conditions.
3. Create a specific **to-do list** which outlines your tasks for the day/week/month/term – include short- and long-term tasks you have to complete.
4. Focus on your **weaknesses** but do not neglect your stronger subjects.
5. Getting help from **tutors** and **teachers** is very useful as well.
6. Go to **bed early** and **wake** up **early**.

Key skills:

1. **Discipline** – you need to constantly be working towards your goals.
2. **Time management** – it is important to keep in touch with extracurriculars which holistically develop you as an individual.
3. **Focus** – you need to train your mind to concentrate for up to three hours at a time. Focus can be learnt by doing anything which requires long concentration like reading or **practice exams**.

A.J. – ATAR 99.25

Five top tips:

1. **Sleep** was very important for me – feeling well rested on exam days made me feel less stressed.
2. **Past papers** were the best study tool – they helped identify weaknesses and are good **practice**.
3. **Balance** helped me – I made sure to do things I enjoyed where I wasn't studying. I took at least a couple of hours off per day.
4. **Breaking up study** into specific topics so I could tackle one at a time helped me to feel like I accomplished something each day (small goals).
5. **Practise** identifying questions that take longer than others – avoiding these until the end made me feel less time stressed during exams.

Key skills:

1. Key skills for success are **time management in exams**. This means realising what questions need more time and leaving them until last to allow for clarity of thought.
2. The other major key skill would be **understanding** the **question** and what the **marker wants**. Doing this well means anything you write will not be wasted. If you're not writing what the marker wants, then often those words are useless and have wasted time for other questions.

T.L. – ATAR 99.95

There are only two 'strategies' I would say I implemented:

1. Ensuring that I was engaging with **classwork** and fully **understood** what was happening.
2. Doing **past papers** as revision and a test for my knowledge before exams.

Key skill:

1. An interest in **learning** and **understanding** the material presented.

N.B. – ATAR 99.1

Top five tips/strategies (in no particular order):

1. **Consistency and repetition**. I found that a daily 30-minute review of my English quotes or syllabus notes three to four times a week was a much more effective method of learning than a huge 2-hour block of memorisation.
2. Build strong academic relationships with your peers and teachers. **Collaboration** is so helpful – cross-reference, share notes, test each other, etc. Your peers are experiencing the same struggles as you, and your teachers are there to help you. Don't be afraid to seek clarification or assistance from your teachers, they really can take your understanding of a topic to the next level.
3. Ensure you are always **prioritising**. Start/focus on the content you find most difficult (within reason).
4. Set **realistic goals** for your study sessions or academic results. Break down your syllabus content into digestible chunks.
5. Practise **wellbeing**: Ensure you are getting enough sleep, exercise, and nutrients. A deficit in any of these areas adversely affects your productivity, and by extension, your results. Don't forget your happiness and health are important as well. As hard as learning new things is, you should try to enjoy the process.

Bonus: **PAST PAPERS ARE ESSENTIAL**. I would contribute a lot of my success to just the vast number of papers I did before every assessment, particularly the trial exams. You can break down these 3-hour papers into 3 x 1-hour study sessions and build towards practising full length papers. Doing past papers under **timed conditions** is the closest you can come to replicating an exam. This helps settle the nerves when HSC comes around.

Key skills:

1. The ability to build and maintain **discipline**. This doesn't mean hours of unwavering concentration and never touching your phone, but ensuring you are able to actually adhere to your goals/tasks every time you sit down to study allows you to maximise productivity. This goes hand in hand with **routine**. It's important to recognise that this is easier said than done, but possible to really develop during trials and HSC. Start small.
2. **Time management**: Again, easier said than done, but use your family, friends, timers, etc. to keep you accountable to studying.
3. Experimenting with different **memory techniques**. I found a mix of handwriting and speaking out loud worked best for me.
4. Ability to **self-reflect**, maintain perspective. I always reminded myself to be realistic. Success can be subjective. I found that by reflecting on this, I was able to push myself to be my best, whatever that may be. Back yourself.
5. **Collaboration**. Working with others can be really helpful when tackling new content, but the real study hours have to be done **individually**. No one else can help you during an exam, it's just you.
6. Condensing and summarising **notes** efficiently and effectively (mind mapping, whiteboarding, compartmentalising).

H.C. – ATAR 99.45

Top five tips and strategies:

1. Spending more time with a **friend** who is **academically driven**. You will push each other to study when others don't.
2. When getting closer to HSC, it's helpful to **study** at the **library during lunchtimes**.
3. For mainly maths and STEM subjects, keep a **master document of questions you got wrong**. Have a section for **past paper** questions and **textbook** questions, copy and paste difficult questions and snip the answers beneath. When getting closer to an exam, just make sure to **redo questions** rather than doing a ton of new ones. This document should be used throughout the year. It helps to colour code: red = I need to do again; yellow = I was close; green = I don't need to redo.
4. Creating **shared documents** with a couple of academic friends is really helpful. Establish that you're both willing to put in effort. Especially for English where teachers teach differently, it can be helpful to see different perspectives.
5. Phone **apps** such as YPT **to log study time**. You make a group with friends and gauge how long you should be studying compared to friends. On a similar note, if you're someone like me who **doomscrolls** on Instagram, during trials/exams, just **delete** it. Reinstall after your last exam.

Some key skills:

1. **Organisation/time management**. Even things like keeping files/emails clean is just one less thing to worry about that you can decide to do at any time. Same thing with your room (or at least your desk). Something helpful if you do STEM subjects could be to decorate your desk:
    1. Annotate your data sheet, e.g. search up an annotated data sheet and copy out the annotations. Even if you haven't

   learnt the formulas yet (some are only used for specific situations), learn those and write them down.
   2. Stick them in front of your desk when you work. Then you don't have to search up data sheets all the time, they'll be right in front of you. Also it looks kind of nice. (You can do this anywhere else too, e.g. English quotes on the other side of the shower door.)
2. **Resilience**. I had some bad tests. Seriously bounce back and acknowledge your mistakes and learn from them.
3. **Kindness**. Be kind to yourself and reward yourself. Don't go cold turkey. Even the week before exams, you can go out for dinner or watch a movie. You've really got to watch out for burnout.

D.W. – ATAR 99.0

Five tips:

1. Always think of **progress in small increments**: Procrastination for me often stemmed from feeling overwhelmed, and motivation came most readily if I felt I was achieving something. I applied this in preparation (i.e. tackling one syllabus question at a time or working solely on a singular skill like handwriting). This sped up progress and made goals seem more achievable.
2. Do **simple tasks** in advance: When I felt tired and unmotivated, I often made sure I had material which I knew I had to work on soon prepared for the next day, e.g. printing out **past papers** and putting them on my desk.
3. Learn answer structures and **how to answer questions**: **Note taking** should focus on more general applicable skills (especially STEM subjects) rather than pure theory. For most information, you can always go back and look through in textbooks, however, minor details like when to round numbers, what key words **markers** look for, specific answer structures for different questions, are often neglected.
4. **Prioritise** subjects: Focus on subjects that will have assessments first and once you have a pretty fleshed-out understanding of your skills don't be afraid to dedicate your time disproportionately for your weaker subjects.
5. **Simulate test conditions** well: Work in **timed conditions without notes**. Make sure to do each test in one go. No music. Occasionally move spots from your preferred study area, as the assessment hall may not be that comfortable. Cut the assessment time you give yourself by 5–10 minutes to account for stress which will inevitably slow you.

Key skills:

1. **Collaboration**. Collaborate intelligently, sometimes if you feel like you are easily distracted it is fine to work alone, but in the

end share information with your peers. Compete on skill, not by withholding information (probably only useful if the people around you have the same mentality).

2. Fast and legible **handwriting**. Make sure your letters are distinctly identifiable from one another. Even if they all look pear shaped, if each one is distinct, handwriting is still legible. If this is an issue, steer away from cursive. Train to write fast and not to tire out. Change pen grips if you have to.
3. **Learn what you can get away with** (chemistry and physics specifically). I often did not write full sentences, instead opting for a 'train of thought' with key words such as 'Disturbance from volume reduction – > Increased pressure – > increased concentration of gases – > Stoichiometric ratio 2:3 – > Reverse reaction favoured to decrease concentration/pressure'. This allowed me to dedicate my time to more difficult questions. Furthermore, the key words became more prominent (all other subjects, except maths proofs, you should write in full sentences).

C.H. – ATAR 99.95

Key tips and strategies:

1. Focusing on **past papers**, even when I didn't feel completely ready.
2. Doing as many **practice questions/essays** as possible and then receiving **feedback.**
3. Having various **study locations** to prevent stagnancy.
4. **Helping/teaching** other peers to consolidate your own knowledge.
5. **Practise without notes** as soon as possible, go back and fill in the gaps once you finish your first attempt.

Key skills:

1. **Discipline and focus,** i.e. being able to focus well for long amounts of time, even when motivation is low.
2. **Collaboration** with peers.
3. **Balance** – having other activities to focus on so studying doesn't feel overwhelming.
4. **Self-reflection** – knowing which specific areas I struggled most with and asking for help.

T.J. – IB 45/45

Key tips and strategies:

1. Making multiple rounds of **notes**
   In class, I mainly focus on **understanding** the content and only take rough notes. Once we finish learning about the topic and I have a better understanding of how different ideas link together, I make a more comprehensive set of summary notes which I use for **revision**.
2. **Past paper mark schemes**
   I read the mark schemes to understand what each mark is rewarded for (e.g. writing down the chain rule formula, attempting to integrate). This helps me structure my answers in a clear way.
3. **Topic tests**
   These help me target topics that I am weaker in. I normally do these after completing full sets of past papers and focus on questions I repeatedly struggle with.
4. **Flashcards**
   Flashcards are really useful for memorising English quotes. I make multiple versions and gradually shrink the quotes down until I can remember them based on one keyword.
5. **Worksheets**
   I make my own worksheets based on my notes. As I read through my notes, I copy them down and leave out the keywords, then I go back and fill in the gaps. This helps me memorise specific details, such as data for geography case studies or material characteristics for DT.

Key skills:

1. **Organisation** and **time management**: **planning** beforehand helped me manage deadlines, especially for projects like the extended essay and DT, which took several months to complete.
2. **Collaborating** with peers: we shared resources (e.g. past papers, PowerPoints) and answered each other's questions.

Key takeaways from these brilliant respondents:

I was tremendously relieved to see that almost all the strategies and skills mentioned were already covered throughout this book. It provided additional validation that the learnings we have shared have been proven to work, having been successfully employed and instrumental in the success of some of the highest achieving HSC students.

When reviewing the responses, it became clear that there were many similarities and common themes regarding their key strategies and skills for academic excellence.

By far the most **common strategies and skills** were:

- Practice questions and past papers – mentioned by every student!
- Discipline and focus – mentioned by almost all students
- Planning, organisation and time management – mentioned by almost all students
- Maintaining a healthy balance – mentioned by half the students
- Chunking work or making lists – mentioned by half the students
- Collaboration – mentioned by half the students

Students also mentioned a variety of other **strategies and skills** such as:

- Doing early and consistent revision
- Getting help and feedback from teachers and tutors
- Prioritising
- Understanding the questions and marking criteria
- Making notes and applying various study and revision techniques
- Goal setting
- Self-reflection
- Getting enough sleep
- Engaging fully with classwork
- Good handwriting
- Pacing yourself

- Resilience
- Using study apps

I want to be clear that although I have only canvassed the opinions of top performing students (because they have unequivocally demonstrated the successful application of good study habits), these skills and strategies are relevant and accessible to most students in any year group who are striving to achieve their personal academic best.

***

Having stood on the sidelines watching countless students, including my own kids, navigate the challenging path through high school and the gruelling final stretch of the HSC, I have gained a clarity of perspective. Beyond the study tips and planning strategies there lie some universal truths. So, with that in mind, I wanted to leave you with some final thoughts …

2. Final thoughts and some universal truths

TRUTH 1

In the absence of motivation – self-discipline is the key!

Throughout high school, especially as you approach the HSC, you will often hear about 'motivation'. You'll be told you are supposed to be self-motivated, driven by passion and curiosity to set and achieve academic goals. Teachers and motivational speakers will give inspirational talks about the necessity of intrinsic motivation. This is wonderful, but the problem with focusing on motivation is that students who don't feel motivated can question whether they are missing a key ingredient for academic success. This self-doubt can unnecessarily impact confidence and performance.

In our experience, motivation can be elusive and even the highest achieving students confess to feeling demotivated some of the time (or even most of the time). This is ok. It is not a failing but a reality, especially for students under pressure. We have learnt that **in the absence of motivation, self-discipline is the key to getting the job done**. The ability to push through and persevere even when you don't feel like it, is a critical and defining factor in academic achievement.

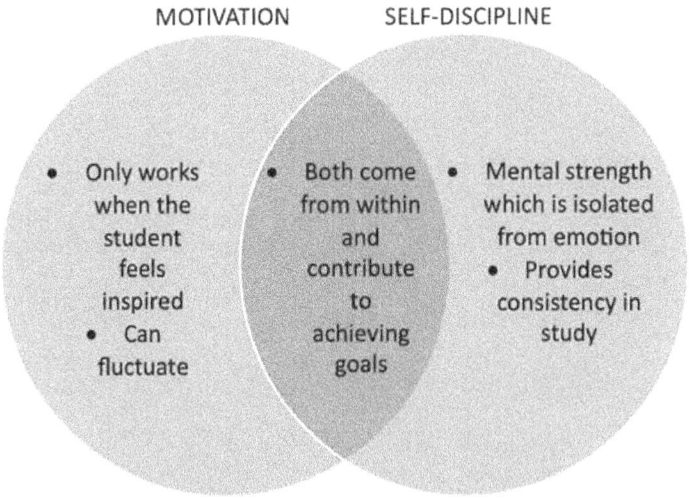

Figure 11.1 Venn diagram – motivation vs self-discipline

In short: while it's ideal to have a combination of motivation and self-discipline, when you feel demotivated, self-discipline is the key to achieving your goals.

> **<u>TRUTH 2</u>**
> **The HSC is a marathon, not a sprint, and consistency is rewarded!**

While this is a cliché, it's also true. Although the pressure intensifies as you move through high school, peaking in Year 12, it's important to maintain perspective and pace yourself. This is where all the tips, techniques, and strategies we have shared will benefit you. If you work consistently in a smart and sustainable way, applying the frameworks and principles we have outlined, you will steadily build mastery, creating optimal conditions for achieving academic excellence.

TRUTH 3

Working collaboratively with your peers will pay dividends!

Collaboration is one of the critical '21st century skills', and an important one to master, but aside from developing teamwork skills, it builds connection, becoming a significant source of academic and emotional support, especially during the HSC.

Studying can be isolating, but working with others who have similar goals is beneficial, creating a supportive and encouraging environment. It can also be of academic benefit as students share resources, study techniques, and assist each other with revision, elevating the performance of the whole cohort, positively impacting final HSC results.

Collaboration does not necessarily require students to work in the same physical space. There will be times when study groups can be counterproductive and individual study is more effective. Regardless of whether students opt for group or solo study, or a combination of both, developing a broader collaboration with fellow students will benefit both the individual and the group.

TRUTH 4

It's vital to look after yourself and maintain a healthy balance and perspective!

Remember: 'This too shall pass', look after yourself, and ask for help if you need it.

Year 12 is very challenging for most students in many respects. It's physically, emotionally, and academically demanding. It can be an overwhelming time of change, and students are often pushed to the limits – overscheduled, sleep deprived, and immensely stressed as they operate within an academic pressure cooker. It may feel relentless and unbearable at times, but try to remember:

- You are not alone – there are many thousands of students feeling the same way
- Year 12 is finite – time passes and you won't feel this way forever
- Whatever the outcome, it does not define your future – there are many paths to get where you want to go
- There are support systems you can tap into for help …

If you are struggling, it's important to **reach out for help** sooner rather than later:

- For **emotional support**, you could reach out to friends, siblings, parents, or other trusted adults, teachers, school counsellors, wellbeing staff, or outside organisations such as 'headspace' and 'Beyond Blue', among others.
- For **academic support,** you could utilise school tutorials and other available resources, reach out to teachers, or engage private tutors. In our experience, most teachers are willing to provide support as they want their students to succeed … and there are a few special teachers who will be prepared to go above and beyond to help students excel. We were lucky enough to have a few of these teachers.

And finally, it's important to look after yourself by:

- Maintaining good sleep, dietary, and exercise habits to keep you physically strong and help with stress management.
- Maintain a healthy balance of social and relaxation activities, spending time with family and friends to enjoy downtime and provide mental respite.

Although we have been talking about this in the context of Year 12, the reminder to maintain perspective, keep healthy and reach out for help applies to students in every year, because for many students the experience of feeling overwhelmed by stress and pressure is not confined to Year 12, and the development of sustainable healthy habits is important at every stage in life.

***

A VERY SHORT CONCLUSION

Having reached the end, I realise that I ended up saying a lot more than originally intended. When I started writing, I thought I would share a few key tips and strategies and that this might only fill twenty or less pages. However, as I wrote, one thing led to another, and it became increasingly apparent just how many different skills and strategies are needed to build towards academic success.

In this study guide, we have covered a lot of ground in a fair amount of detail – talking about everything from the transition to Year 7 right up to the end of the HSC exams. The focus throughout has been to provide tips and strategies applicable to all year groups, many of which are not explicitly taught in the classroom, but all of which are crucial to achieve sustained academic excellence. The fundamental principles and value of mindset, consistent practice, working smart, prioritising, managing time and resources, applying the 80/20 rule, setting goals, implementing day-to-day strategies, study planning, assessment preparation, and developing the skills and techniques to conquer exams … are **all** an intrinsic part of achievement. While these strategies are helpful for most students, making life easier and improving performance, they are absolutely essential for students with high academic aspirations. It would be very difficult to excel academically without consistently employing most of these techniques. With all the best intentions, I understand that telling students **what** they should ideally do isn't particularly helpful if they aren't given guidance on **how** to do it. I have therefore tried as far as possible to provide practical and actionable tips and examples that students can implement immediately to enhance their academic performance.

This book took me longer than expected to write, and in a way, it has been advantageous, because most of the content was written while my daughter was navigating her HSC year. This confluence of timing, while challenging, brought everything into much sharper focus, as I have been (even more) acutely aware of the necessity of each and every skill and strategy, up close and in real time. It has also served as a reminder of just how tough high school can be, especially at the pointy end when students are overwhelmed, exhausted, stressed, and under immense pressure. This reinforced just how important it is to develop great study skills and habits early on, because if they aren't entrenched by the time a student reaches Year 12, things are harder than they need to be.

I set out with the goal of sharing learnings that we acquired over the years through personal experience and observation of other high-achieving students. I hoped that these might ease the path for other students, remove some academic obstacles, alleviate the stress that comes from figuring things out the hard way, and assist them in their pursuit of excellence and achievement of their highest academic aspiration. I hope I have achieved this goal.

Thank you for reading ☺

ACKNOWLEDGEMENTS

There are several people that were instrumental in the writing of this book, to whom I owe a debt of gratitude.

First and foremost, I am tremendously grateful to our amazing children, Alex and Bella. Obviously without you this book could never have been written. Thank you for allowing me to be part of your educational journeys and for your willingness to share all your learnings with me over the years. Thank you for giving me access and allowing me to include excerpts of your personal work and study plans, as well as permitting me to share some of your hard-learned lessons and a few painful experiences. I appreciate enormously your investment of time and energy in reading my manuscript, providing insightful feedback and unwavering support. You are the only reason this book exists. Beyond your outstanding academic achievements and valuable contribution to my literary endeavour, I am immensely proud of the incredible humans that you are. I feel truly blessed to be your mom. I love you both to infinity and back.

Thank you to my incredible husband, Christos! You have supported and encouraged me from the very beginning. You pushed me to persevere even when I was filled with fear and self-doubt. Thank you for being my sounding board, brainstorming with me, and sharing your business perspective. B2B!

Thank you to Sue and Le'Roy, our long-time friends. You are the ones who ignited the initial spark for this project … with a casual comment made over a cup of coffee on a Saturday afternoon. You have jointly given us the benefit of your wise perspectives and lateral thinking on many issues over the years for which we are ever grateful … and Sue, for more than forty years, you have been a great friend, sounding board, and unconditional support.

Thank you to Tania, one of the busiest people I know, for going beyond any reasonable expectation and taking on the mammoth task of reading my unedited manuscript, which was not exactly a quick or light read. Beyond your extremely valuable feedback on the contents of this book, I am enormously grateful to you for sharing your business perspective and workshopping various strategies and scenarios with me beyond the publishing phase.

I will be forever grateful to my mom (Carol) and dad (Ken) for many things – among them, their unconditional love and support, and their investment in my education – instilling in me an early appreciation for the value of learning and the doors that it opens. I am who I am because of the way I was raised and the example they both set for me. Their virtues are numerous, but in the context of education they taught me the importance of work ethic, self-discipline, striving to be your best, and persevering in the face of challenge. These values are infused throughout this book as I believe they are foundational to achievement, both at school and in life.

Thank you to the high school graduates who were willing to engage with my research, taking the time to provide their crucial insights, and permitting me to publish their responses and ATAR results (I have used initials in order to protect the identity of the students): A.A., I.A., N.B., H.C., C.H., A.J., T.J., T.L., S.R., D.W.

I am also indebted to the many inspirational teachers and educators that we were lucky enough to encounter at various points along the way.

Thank you to Ejean for your friendship and support, which started way back when our oldest kids were in Year 3. Our weekly coffees, which kept me sane, were largely dedicated to mutually lamenting the hard lessons our kids had to learn along their educational journeys – some were humorous, others not so much, and many have made it into this book in one way or another.

Thank you to the friends (the very few that I told) for cheering me on and encouraging me quietly from the sidelines.

Thank you to Lesley for your expert editing and proofreading, kind words of encouragement, and patience with my many questions and emphasis quirks.

Thank you to Andy and the team at Publicious for your guidance, support, and help with so many aspects of the publishing process that I would not have managed to navigate on my own.

BIBLIOGRAPHY AND FURTHER READING

Australian Government. (2026). *24-hour movement guidelines for all Australians.* Australian Government Department of Health, Disability and Ageing. https://www.health.gov.au/topics/physical-activity/24-hour-movement-guidelines-for-all-australians

Australian Government Department of Health. (2023). *Australian Government Department of Health.* https://www.health.gov.au

BBC. (n.d.). *What are plant and animal cells?* BBC Bitesize. https://www.bbc.co.uk/bitesize/articles/zkm7wnb#zxnvf82

Beyond Blue (https://www.beyondblue.org.au)

Burns, R., Fu, Y., Brusseau, T., Clements-Nolle, K., and Yang, W. (2018). Relationships among physical activity, sleep duration, diet, and academic achievement in a sample of adolescents. *Preventive Medicine Reports, 12,* 71–74. https://doi.org/10.1016/j.pmedr.2018.08.014

Campion, J. (Director). (2009). *Bright Star.* Jan Chapman Productions.

Cirillo, F. (n.d.). *The Pomodoro® Technique.* pomodorotechnique.com. https://www.pomodorotechnique.com

Claremont Graduate University. (2014). *Mihaly Csikszentmihalyi* [the "father of flow"]. Claremont Graduate University. https://www.cgu.edu/people/mihaly-csikszentmihalyi/

Deakin University. (2020, April 16). *The great big multitasking myth: why it makes you less efficient.* https://this.deakin.edu.au/self-improvement/the-great-big-multitasking-myth-why-it-makes-you-less-efficient/

Dickens, C. (2003). *Great Expectations.* (Mitchell, C. ed.) Penguin Classics. (Original work published 1860.)

Doran, G. (1981). There's a S.M.A.R.T. way to write management's goals and objectives. *Journal of Management Review, 70*, 35–36.

Dweck, C. (2006). Mindset: How you can fulfil your potential. Constable & Robinson.

Ebbinghaus, H. (1913). *Memory: A contribution to experimental psychology* (H.A. Ruger & C.E. Bussenius, Trans.). New York, NY: Teachers College, Columbia University. (Original work published 1885)

Eisenhower, D. (1954, August). *Address at the Second Assembly of the World Council of Churches, Evanston, Illinois | The American Presidency Project.* Ucsb.edu. https://www.presidency.ucsb.edu/documents/address-the-second-assembly-the-world-council-churches-evanston-illinois

Encyclopædia Britannica. (2019). Hermann Ebbinghaus | German psychologist. In *Encyclopaedia Britannica.* https://www.britannica.com/biography/Hermann-Ebbinghaus

Evans, B. (2012). Editor's perspective article: nutrition, exercise, and sleep: Physiological considerations in the classroom for alternative certification teachers. *JNAAC, 7*(1). https://files.eric.ed.gov/fulltext/EJ1053326.pdf

Faught, E. L., Ekwaru, J. P., Gleddie, D., Storey, K. E., Asbridge, M., & Veugelers, P. J. (2017). The combined impact of diet, physical activity, sleep and screen time on academic achievement: a prospective study of elementary school students in Nova Scotia, Canada. *International Journal of Behavioral Nutrition and Physical Activity, 14*(1). https://doi.org/10.1186/s12966-017-0476-0

Fullan, M., & Scott, G. (2014). *New Pedagogies for Deep Learning Whitepaper: Education PLUS The world will be led by people you can count on, including you!* https://www.michaelfullan.ca/wp-content/uploads/2014/09/Education-Plus-A-Whitepaper-July-2014-1.pdf

Gladwell, M. (2008). *Outliers: The story of success.* Penguin.

Headspace (https://headspace.org.au)

Icez (original uploader at English Wikipedia). (2007). *A typical representation of the forgetting curve.* Wikimedia Commons. (Figure adapted from ForgettingCurve.svg, Wikimedia Commons.)

Joshi, A., & Hinkley, T. (2021). *Too much time on screens? Screen time effects and guidelines for children and young people.* Australian Institute of Family Studies. https://aifs.gov.au/resources/short-articles/too-much-time-screens

Kruse, K. (2016, March 7). *The 80/20 rule and how it can change your life.* Forbes. https://www.forbes.com/sites/kevinkruse/2016/03/07/80-20-rule/?sh=7fca55eb3814

National Health and Medical Research Council. (2013). *Recommended number of serves for children, adolescents and toddlers | eat for health.* Eatforhealth.gov.au; Australian Government. https://www.eatforhealth.gov.au/food-essentials/how-much-do-we-need-each-day/recommended-number-serves-children-adolescents-and-toddlers

NESA. (2022, September 15). *NESA | NSW Government.*nsw.gov.au. https://www.nsw.gov.au/education-and-training/nesa

NSW Education Standards Authority. (2023). *Glossary of key words | NSW Government.*nsw.gov.au. https://www.nsw.gov.au/education-and-training/nesa/hsc/student-guide/glossary

NSW Education Standards Authority. (2024a). *Summative assessment.* Curriculum.nsw.edu.au. https://curriculum.nsw.edu.au/assessment-and-reporting/summative-assessment

NSW Education Standards Authority. (2024b, September 10). *2025 HSC enrolment snapshot.* NSW Government. https://www.nsw.gov.au/education-and-training/nesa/hsc/facts-and-figures/enrolment-data

NSW Education Standards Authority. (2024c, November 21). *11 – 12 English syllabuses.* NSW Government. https://www.nsw.gov.au/education-and-training/nesa/curriculum/english

NSW Education Standards Authority. (2025, April 28). *English Advanced HSC exam papers.* NSW Government. https://www.nsw.gov.au/education-and-training/nesa/curriculum/hsc-exam-papers/english-advanced

Olson, E. (2023, February 21). *How many hours of sleep do you need?* Mayo Clinic. https://www.mayoclinic.org/healthy-lifestyle/adult-health/expert-answers/how-many-hours-of-sleep-are-enough/faq-20057898

Percival, B. (Director). (2013). *The Book Thief.* Fox 2000 Pictures; Sunswept Entertainment; Babelsberg Film

Shakespeare, W., & Smith, R. (2014). *The Merchant of Venice.* Cambridge University Press.

Shakespeare, W., Smith, R., Gibson, R., Wienand, V., & Andrews, R. (2014). *Romeo and Juliet.* Cambridge University Press.

Shakespeare, W., & Spencer, R. (2014). *Twelfth night.* Cambridge University Press.

Sutton, J. (2016, October 4). *Martin Seligman's positive psychology theory.* PositivePsychology.com. https://positivepsychology.com/positive-psychology-theory/

THSC Online (https://thsconline.github.io/s/)

Zusak, M. (2005). *The Book Thief.* Black Swan.

NOTES

1. I would like to share a little more information to support my rationale and substantiate the effectiveness of the methods and techniques proposed in this book. Through the consistent application of the strategies outlined in these pages, both our kids were able to achieve academic honour awards every year of high school, top a variety of subjects across both STEM and humanities, while holding leadership positions. Additionally, both were invited to, and joined, the accelerated mathematics programmes at their respective schools, and both were consistently 'A' students. Our son was awarded a full academic scholarship for Years 11–12, earned an ATAR of 99.5, was a Distinguished Achiever in all his 12 units and an All-round Achiever, receiving the Premier's Award for All-round Excellence in the NSW HSC in 2022. He is currently studying a Bachelor of Quantum Engineering (Hons) and a Bachelor of Science double degree at UNSW. Our daughter earned an ATAR of 99.8, was a Distinguished Achiever in all her 12 units, an All-round Achiever and a state ranker, placing 10th in the state for modern history. She also received the Premier's Award for All-round Excellence in the NSW HSC in 2025, and she is aspiring to study medicine. These results would not have been possible without diligent application of the skills and strategies we have shared.

2. The research I conducted among high-achieving students in the HSC/IB was done via questionnaire with open-ended questions centred on their opinions of the top skills and strategies required for high academic performance. The ten students were all selected based on the criterion of an ATAR over 99. The sample included both male and female respondents with diverse cultural backgrounds who graduated between 2020 and 2025, from a selection of schools.

3. The included excerpts of the diary and study plans were created by my children for their own personal study and have been used in this book for illustrative purposes, with their permission.

INDEX

80/20 rule 22, 24, 27, 40, 139, 266, 304, 317, 318, 323, 377, 385

A

acceleration 79, 343, 344, 345

adapt 136, 158, 168, 202, 204, 207, 241, 242, 246, 247, 251, 252, 253, 254, 255, 257, 263, 264, 265, 266, 268, 357

aesthetics 188, 260, 261

analyse 100, 102, 103, 104, 109, 110, 119, 120, 128, 129, 131, 134, 137, 138, 157, 165, 178, 230, 231, 239, 240, 243, 244, 260

answering the question/ATQ 100, 123, 124, 125, 126, 128, 132, 133, 135, 136, 144, 147, 156, 157, 162, 163, 164, 167, 171, 172, 174, 175, 176, 177, 178, 179, 183, 198, 245, 251, 252, 253, 255, 256, 266, 357

apply i, 19, 34, 39, 45, 50, 51, 73, 80, 102, 105, 110, 111, 120, 121, 128, 129, 134, 135, 143, 148, 151, 155, 165, 168, 172, 174, 198, 202, 206, 207, 214, 221, 222, 223, 233, 265, 278, 284, 289, 292, 295, 326, 343, 358

assess 30, 59, 81, 100, 102, 104, 114, 117, 119, 120, 121, 122, 131, 134, 138, 153, 253, 322

assessment notification/notification viii, 60, 84, 86, 87, 89, 90, 91, 96, 98, 99, 123, 140, 148, 153, 158, 162, 185, 192, 228, 243, 272, 299, 306, 308, 309, 310, 311, 312, 324

assessment/exam preparation viii, 4, 15, 25, 26, 28, 37, 46, 55, 59, 65, 69, 77, 78, 79, 81, 82, 83, 84, 85, 86, 87, 89, 90, 91, 93, 95, 96, 98, 99, 100, 105, 114, 115, 121, 123, 128, 134, 139, 140, 141, 142, 143, 146, 147, 148, 153, 158, 162, 165, 179, 185, 186, 188, 190, 192, 196, 198, 202, 203, 204, 209, 210, 225, 228, 229, 243, 248, 255, 270, 271, 272, 274, 280, 284, 285, 289, 290, 291, 294, 297, 298, 299, 300, 303, 304, 305, 306, 308, 309, 310, 311, 312, 313, 315, 316, 317, 319, 320, 321, 322, 323, 324, 333, 335, 347, 350, 356, 363, 367, 377, 386

ATAR i, iii, ix, xi, 22, 75, 77, 148, 152, 301, 325, 329, 330, 339, 340, 341, 342, 344, 353, 355, 356, 357, 358, 359, 360, 361, 362, 363, 365, 367, 369, 380, 389, 390

B

balance iii, 24, 26, 63, 68, 133, 188, 189, 316, 322, 325, 331, 356, 361, 369, 371, 374, 376

C

calculator 8, 9, 85, 89, 211, 223, 229, 278, 279
Carol Dweck 16, 17, 19
characters 109, 158, 159, 160, 165, 166, 170, 172, 230, 231, 235, 237, 238, 240, 244, 245, 246, 262, 310
classwork 40, 45, 47, 48, 51, 56, 98, 190, 193, 207, 271, 272, 298, 299, 309, 362, 371
collaboration 71, 83, 97, 290, 291, 363, 364, 367, 369, 371, 374
compare 100, 102, 103, 104, 111, 112, 113, 119, 120, 135, 138, 141, 224, 234, 287, 365
competitions 69
composer 116, 161, 178, 196, 235, 236, 237, 238, 240, 242
conceptual 51, 125, 130, 140, 156, 160, 161, 162, 163, 164, 166, 167, 175, 189, 196, 224, 233, 237, 243, 245, 247, 251, 266, 271, 293
conclusion 15, 32, 67, 102, 109, 118, 138, 144, 157, 158, 178, 179, 180, 247, 248, 251, 253, 287, 290, 377
consistency 39, 152, 326, 357, 363, 373
context iii, iv, vii, viii, 29, 88, 100, 105, 107, 119, 149, 150, 157, 169, 177, 178, 196, 230, 231, 232, 235, 236, 240, 243, 245, 251, 310, 376, 380
contrast 47, 102, 104, 111, 112, 113, 311
conventions 150, 156, 174, 178, 244, 256, 258, 260, 265
creative 70, 78, 82, 87, 131, 147, 227, 233, 256, 257, 258, 259, 260, 264, 265, 313
Csikszentmihalyi 66, 67, 384
curriculum vi, 42, 71, 77, 78, 79, 82, 83, 94, 189, 190, 192, 193, 209, 229, 230, 234, 298, 343, 349, 350, 386

D

describe 1, 66, 100, 101, 102, 103, 104, 106, 109, 114, 116, 137, 145,
diagrams 188, 190, 201, 223, 283
diary 52, 53, 54, 55, 56, 59, 63, 304, 305, 306, 307, 312, 390
diminishing returns 21, 22, 23, 24, 27, 31, 34, 50, 152, 188, 206, 208, 226, 248

directive terms 103, 104, 105, 117, 118, 119, 120, 121, 122, 124, 128, 130, 132, 133, 144, 145, 163, 165, 176, 223, 226, 227, 228, 229, 252, 269, 284
disappointment 17, 291
discursive 256, 257, 259, 260, 264,
discuss viii, 36, 87, 89, 102, 104, 107, 108, 109, 335, 337
distractions 33, 57, 58, 211, 217, 332, 359, 360

E
Ebbinghaus forgetting curve 199, 200, 384
Eisenhower 26, 384
electives 2, 78, 335, 336
English i, vi, 8, 24, 25, 26, 28, 33, 46, 50, 53, 55, 67, 78, 79, 87, 94, 98, 104, 108, 109, 111, 114, 115, 116, 117, 119, 121, 125, 130, 131, 134, 135, 137, 148, 149, 150, 151, 152, 153, 157, 158, 159, 162, 163, 168, 171, 175, 177, 179, 180, 182, 193, 194, 195, 196, 204, 214, 215, 218, 229, 230, 231, 232, 233, 234, 240, 241, 242, 243, 248, 256, 261, 266, 267, 268, 269, 270, 280, 281, 283, 285, 294, 296, 299, 310, 311, 314, 317, 319, 325, 331, 335, 337, 338, 340, 341, 342, 346, 347, 350, 357, 363, 365, 366, 370, 385, 386
equation 89, 106, 197, 198, 278
error pattern 221, 225, 227, 292
essay 33, 66, 67, 86, 97, 99, 103, 133, 137, 138, 139, 143, 144, 145, 151, 153, 155, 156, 157, 158, 159, 161, 162, 168, 169, 170, 171, 172, 173, 174, 175, 177, 178, 179, 180, 182, 183, 214, 215, 217, 218, 228, 235, 236, 237, 238, 240, 241, 242, 243, 244, 245, 246, 247, 248, 249, 250, 251, 252, 253, 254, 255, 266, 267, 268, 281, 283, 294, 299, 305, 307, 310, 314, 317, 325, 370
essay structure 144
evaluate 35, 37, 102, 104, 114, 115, 116, 117, 147, 170, 220, 273, 289, 314, 321, 322, 331, 332
evidence 16, 17, 97, 102, 108, 109, 115, 116, 118, 120, 133, 135, 137, 144, 145, 164, 165, 168, 173, 174, 175, 176, 177, 178, 179, 180, 195, 236, 237, 238, 239, 240, 242, 243, 244, 253, 254, 255, 256, 260, 263, 275, 285, 310
exam techniques 185, 229, 270, 274, 280, 282, 286, 295, 300, 351
explain 16, 36, 43, 90, 100, 101, 102, 103, 104, 105, 106, 107, 111, 120, 128, 129, 134, 137, 143, 145, 153, 154, 199, 233, 313
explanation 80, 90, 107, 115, 118, 134, 144, 174, 175, 196, 281
extra units 341, 342

F
feedback 17, 45, 47, 69, 80, 86, 87, 94, 95, 124, 147, 148, 154, 155, 156, 159, 182, 218, 227, 228, 248, 249, 258, 265, 294, 310, 314, 326, 348, 349, 350, 358, 369, 371, 379, 380, 395
film 109, 110, 111, 162, 164, 176, 194, 195, 196, 230, 235, 240, 386
flashcards 60, 188, 201, 370
flow charts 108
flow/flow theory 49, 66, 67, 68, 145, 163, 167, 174, 175, 176, 177, 188, 213, 239, 260, 266, 268, 284, 285, 317, 331, 384
focus iii, iv, 2, 17, 22, 26, 27, 29, 31, 32, 33, 35, 50, 58, 59, 63, 66, 67, 68, 72, 80, 93, 109, 112, 126, 127, 133, 140, 147, 153, 156, 169, 180, 189, 202, 212, 213, 216, 227, 230, 232, 233, 234, 235, 238, 241, 242, 243, 244, 245, 246, 247, 248, 249, 250, 252, 254, 255, 256, 258, 260, 262, 265, 266, 271, 274, 275, 282, 282, 291, 310, 314, 316, 317, 322, 325, 331, 341, 343, 350, 359, 360, 363, 367, 369, 370, 371, 377, 378
focus area 232, 233, 234, 235, 241, 242, 243, 244, 245, 246, 247, 248,249, 250, 252, 255, 256, 258, 260, 265, 310
form i, 21, 41, 46, 82, 88, 97, 108, 149, 162, 163, 165, 166, 170, 172, 173, 187, 196, 235, 236, 240, 243, 244, 245, 251, 256, 257, 258, 259, 273, 294, 310
formula 167, 168, 180, 189, 222, 277, 370

G
goals iv, xi, 3, 13, 14, 35, 36, 38, 63, 67, 289, 292, 295, 299, 311, 314, 316, 331, 346, 355, 360, 361, 363, 364, 367, 372, 373, 374, 377, 384

H
handwriting 47, 182, 183, 187, 217, 222, 279, 284, 286, 364, 367, 368, 371
homework 2, 8, 15, 24, 25, 26, 28, 42, 43, 44, 45, 46, 47, 48, 49, 50, 51, 52, 53, 55, 56, 58, 59, 62, 64, 69, 100, 183, 192, 230, 271, 272, 298, 299, 309, 312, 315, 316, 326, 341, 345, 348, 349
HSC ii, vi, vii, ix, 8, 18, 22, 39, 78, 82, 94, 101, 118, 121, 150, 153, 156, 158, 175, 179, 182, 187, 206, 209, 214, 215, 229, 230, 231, 233, 234, 235, 241, 243, 248, 256, 263, 267, 269, 270, 272, 280, 282, 283, 291, 295, 296, 300, 303, 305, 312, 317, 322, 323, 324, 325, 326, 327, 328, 329, 330, 331, 333, 336, 337, 338, 340, 342, 343, 344, 345, 353, 355, 356, 357, 358, 363, 364, 365, 371, 372, 373, 374, 377, 378, 386, 389, 390

I
ideas v, 16, 17, 18, 67, 112, 130, 144, 161, 162, 163, 164, 167, 171,

181, 234, 235, 236, 237, 239, 240, 243, 244, 245, 255, 257, 258, 259, 261, 262, 263, 264, 265, 266, 310, 370
identify 17, 24, 30, 38, 82, 95, 102, 103, 104, 108, 113, 117, 122, 125, 128, 129, 130, 136, 138, 173, 193, 212, 218, 221, 228, 229, 251, 252, 280, 289, 292, 307, 343, 361
imagery 115, 116, 239, 240, 259, 260, 261, 262
imaginative 256, 257, 259, 260, 262, 264
importance 14, 26, 27, 29, 30, 31, 32, 35, 49, 58, 59, 67, 122, 155, 158, 159, 162, 163, 171, 203, 204, 226, 236, 292, 303, 305, 315, 317, 320, 323, 380
independent learning 61, 65, 68,
insight 96, 103, 110, 119, 120, 152, 155, 159, 160, 161, 162, 163, 164, 169, 175, 177, 183, 208, 246, 247
instructions 4, 84, 85,90, 99, 127, 129, 130, 224, 227, 228, 269, 284, 286
introduction i, 137, 144, 157, 158, 161, 169, 170, 171, 172, 173, 174, 175, 178, 236, 238, 247, 248, 251, 253, 259

J

justify 102, 104, 118, 119, 127, 128, 224, 260, 284

K

key message 167, 196, 235, 237, 238, 239, 244, 245, 246, 310

L

leadership iii, 10, 25, 46, 70, 296, 389
linking words 107, 111, 112, 113, 118, 129, 163, 165, 176
logic 223, 313, 314, 320, 321, 322, 331

M

macro elements 235, 236, 244, 247, 255
Malcolm Gladwell 14, 17, 203
mark allocation 112, 133, 135, 137, 138, 141, 147, 154, 174, 177, 207, 214, 275, 276, 282
marker 6, 86, 94, 95, 100, 112, 113, 116, 117, 123, 136, 137, 140, 144, 146, 147, 154, 155, 156, 159, 161, 163, 168, 172, 173, 177, 180, 181, 182, 183, 185, 228, 258, 260, 263, 275, 279, 284, 361
marking criteria 86, 87, 88, 90, 95, 140, 141, 142, 143, 144, 145, 146, 147, 148, 149, 150, 152, 153, 154, 155, 156, 168, 173, 180, 181, 183, 185, 227, 228, 233, 243, 247, 269, 284, 293, 307, 310, 371
marking guidelines 86, 121, 148, 154, 155, 156, 157, 208, 218, 293
master 25, 121, 125, 148, 158, 168, 174, 185, 215, 241, 242, 243, 244,

245, 246, 247, 248, 249, 250, 251, 252, 253, 254, 255, 256, 257, 258, 264, 265, 266, 267, 268, 282, 294, 299, 310, 311, 317, 320, 325, 365, 374
math/maths/mathematics i, 7, 8, 9, 16, 25, 26, 28, 37, 42, 47, 48, 50, 51, 53, 58, 59, 66, 69, 78, 79, 83, 85, 86, 89, 98, 104, 127, 138, 155, 182, 196, 197, 203, 204, 207, 215, 218, 221, 222, 224, 225, 229, 269, 270, 271, 272, 273, 275, 276, 278, 279, 280, 281, 289, 296, 308, 309, 310, 311, 314, 317, 321, 325 , 326, 335, 337, 338, 340, 343, 344, 345, 346, 347, 351, 357, 365, 368, 389
memorise/memory 13, 97, 187, 195, 199, 200, 202, 219, 228, 238, 241, 242, 246, 250, 251, 253, 263, 266, 267, 272, 310, 317, 325, 328, 364, 370, 384
micro analysis 235, 236, 245, 235, 236
mind maps 188, 201
mistakes 18, 32, 183, 216, 220, 221, 224, 225, 228, 271, 272, 275, 276, 277, 279, 288, 289, 292, 293, 294, 309, 356, 258, 366
modules 83, 233, 234, 336, 337
motif 195, 239, 240, 259, 263, 264
multi-modal 82, 139, 162, 227, 235,
multi-tasking 32, 33, 34, 42, 59
multiple-choice 207, 223, 275, 276, 280, 281, 282

N
NESA vi, 8, 81, 85, 101, 156, 193, 209, 233, 234, 241, 272, 310, 340, 386
notes vi, 4, 8, 9, 21, 40, 41, 42, 43, 44, 60, 69, 87, 89, 94, 97, 103, 186, 187, 188, 189, 190, 191, 192, 193, 194, 196, 197, 198, 199, 200, 201, 202, 203, 206, 211, 212, 219, 228, 238, 246, 265, 271, 272, 273, 284, 293, 295, 297, 298, 299, 308, 309, 310, 311, 314, 317, 320, 321, 324, 357, 363, 364, 367, 269, 370, 371, 389

O
organisation v, 52, 56, 63, 64, 304, 365, 370, 371

P
paragraph/s 24, 50, 103, 108, 109, 133, 137, 138, 144, 156, 157, 158, 159, 160, 161, 168, 169, 170, 171, 172, 173, 174, 175, 176, 177, 178, 179, 180, 236, 237, 238, 239, 243, 244, 245, 248, 250, 253, 255, 259, 260, 265, 268, 310
past papers 21, 85, 89, 156, 206, 207, 208, 209, 210, 211, 219, 221, 228, 252, 254, 255, 272, 273, 274, 309, 310, 311, 317, 318, 321, 323, 325, 326, 331, 357, 361, 362, 363, 367, 369, 370, 371
PEEL 175, 177
persuasive 256, 257, 260

PETAL 177
plan/planning iv, viii, 13, 25, 28, 36, 38, 47, 49, 50, 52, 54, 56, 60, 61, 62, 63, 65, 66, 67, 68, 78, 85, 93, 109, 171, 179, 191, 204, 212, 268, 273, 284, 294, 300, 301, 303, 304, 305, 306, 307, 308, 310, 312, 313, 314, 315, 316, 317, 318, 319, 320, 321, 322, 323, 324, 325, 326, 327, 328, 329, 330, 331, 332, 333, 348, 356, 357, 370, 371, 372, 377, 395
Pomodoro Technique 65, 383
practice/practise i, iv, 14, 15, 17, 18, 20, 21, 25, 26, 28, 30, 39, 42, 45, 48, 50, 55, 68, 71, 75, 85, 87, 97, 121, 124, 140, 148, 151, 160, 187, 197, 203, 204, 205, 206, 207, 208, 209, 210, 212, 213, 214, 216, 217, 218, 219, 220, 221, 224, 228, 232, 241, 249, 250, 252, 253, 254, 255, 265, 266, 270, 271, 272, 273, 274, 279, 292, 297, 310, 311, 312, 314, 317, 318, 322, 325, 326, 336, 344, 347, 350, 356, 357, 360, 361, 363, 369, 371, 377
presentation 24, 42, 82, 83, 84, 90, 96, 97, 99, 139, 162, 226, 227, 235, 261
prioritising/priority 23, 24, 26, 27, 28, 29, 30, 31, 32, 34, 39, 43, 48, 49, 50, 51, 52, 55, 61, 63, 65, 69, 85, 91, 192, 204, 226, 272, 286, 294, 304, 305, 306, 307, 308, 313, 314, 317, 318, 319, 320, 325, 331, 359, 363, 371, 377
procrastination 29, 49, 66, 307, 332, 358, 367
purpose 26, 40, 67, 68, 86, 88, 103, 115, 149, 161, 178, 196, 198, 234, 235, 236, 237, 238, 239, 240, 242, 243, 244, 245, 246, 259, 260, 263, 310

Q

quadrants 27, 28, 33, 59
quote bank 194, 195, 196, 238, 246, 249, 251, 255, 266, 310, 317

R

ranking 289, 290, 296
reading time 84, 274, 275, 280, 281, 282
reflection 83, 155, 220, 221, 256, 257, 293, 294, 369, 371
reports 5, 45, 46, 71, 98, 139, 228, 289, 383
revision 48, 55, 60, 86, 89, 97, 187, 190, 197, 198, 201, 203, 205, 206, 207, 208, 221, 228, 271, 272, 274, 293, 297, 298, 309, 311, 313, 317, 318, 320, 321, 322, 325, 331, 347, 357, 362, 370, 371, 374
rubric 140, 141, 142, 143, 144, 145, 147, 148, 149, 151, 152, 269

S

S.T.E.P. 313, 331
scaling 338, 339, 340, 341
scene 109, 162, 165, 176, 194, 195, 240

screentime 71, 72
self-discipline 39, 60, 68, 140, 299, 332, 353, 358, 359, 372, 373, 380
Seligman 66, 67, 387
set work 53, 131, 134, 176, 193, 194, 230, 232, 233, 234, 235, 239, 249, 252, 299, 310, 350
skeleton 171, 254, 255, 257, 268, 284, 310
sleep 57, 65, 71, 72, 201, 361, 363, 371, 375, 376, 383, 384, 385, 386
smart priority grid 27, 28, 31, 32, 49, 52, 61, 91
source 54, 114, 115, 120, 131, 132, 133, 134, 204, 207, 208, 265, 276, 279, 281, 314, 329, 330, 347, 348, 374
spaced repetition 199
specificity 313, 314, 331
speech 13, 47, 82, 96, 139, 151, 226, 227, 235
stimulus 123, 131, 132, 133, 134, 135, 253, 257
study apps 332, 333, 357, 372
study plan 308, 312, 314, 317, 319, 320, 324, 325, 329, 330, 332
subject choices 335, 338, 339
subjects 1, 2, 8, 9, 47, 49, 51, 55, 60, 62, 67, 68, 77, 78, 79, 81, 90, 94, 95, 101, 104, 109, 111, 114, 125, 142, 152, 156, 157, 168, 180, 182, 185, 186, 187, 191, 193, 196, 197, 198, 204, 205, 214, 218, 221, 223, 225, 229, 267, 268, 269, 271, 272, 280, 283, 284, 285, 294, 296, 297, 298, 299, 311, 319, 320, 321, 322, 324, 325, 326, 331, 335, 336, 337, 338, 339, 342, 343, 344, 345, 347, 350, 351, 357, 358, 360, 365, 367, 368, 389
subtask 307, 308, 315
synthesise 102, 104, 119, 120, 132

T

task list 31, 311
technique 65, 104, 118, 126, 164, 176, 177, 185, 186, 194, 195, 207, 220, 225, 238, 239, 240, 252, 350, 383
TEEL 175
textbook 7, 8, 20, 21, 58, 85, 89, 98, 186, 188, 190, 191, 197, 199, 205, 272, 309, 317, 365
themes 158, 161, 162, 230, 231, 232, 235, 236, 237, 238, 239, 240, 243, 246, 255, 263, 264, 310, 371
thesis 157, 158, 159, 160, 161, 162, 163, 164, 165, 166, 167, 168, 169, 170, 171, 172, 173, 174, 175, 176, 177, 178, 183, 237, 238, 241, 242, 243, 244, 245, 246, 250, 251, 253, 254, 255, 260, 266, 268, 310
Time management 24, 30, 32, 34, 36, 38, 39, 56, 65, 68, 69, 139, 204, 207, 213, 214, 215, 216, 266, 280, 282, 304, 308, 313, 314, 315, 316, 320, 323, 331, 345, 358, 360, 361, 364, 365, 370, 371
timestamp 194, 195

topic sentence 144, 171, 174, 175, 176, 255, 256
topics viii, 14, 50, 79, 83, 89, 98, 207, 263, 274, 346, 361, 370, 383

traps 31, 126, 188, 222, 223, 224, 225, 228, 271, 272, 273, 277, 282, 293, 309
trials 250, 267, 291, 305, 312, 323, 324, 326, 327, 328, 364, 365
tutor/tutoring 1, 218, 313, 316, 331, 346, 347, 348, 349, 350, 351

U
urgency 26, 27, 29, 30, 32, 49, 55, 59, 314, 317, 319, 320

V
values 88, 128, 149, 150, 178, 196, 223, 235, 236, 237, 240, 243, 245, 277, 310, 380

W
weightings 85, 289, 305, 319
working iii, v, xi, 14, 15, 21, 22, 23, 28, 29, 33, 34, 40, 43, 49, 56, 60, 61, 63, 66, 67, 68, 77, 90, 91, 98, 122, 123, 124, 137, 138, 146, 147, 155, 156, 197, 208, 211, 214, 218, 219, 220, 222, 241, 242, 248, 249, 255, 268, 270, 275, 276, 277, 278, 279, 281, 292, 294, 299, 303, 304, 307, 313, 320, 321, 342, 347, 350, 356, 360, 364, 367, 374, 377
working smart 21, 23, 28, 33, 40, 43, 56, 60, 61, 63, 77, 90, 91, 98, 123, 197, 208, 220, 242, 270, 377

Y
Year 11 ix, 24, 37, 81, 101, 163, 165, 182, 295, 296, 297, 299, 324, 336, 337, 338, 345, 346

Year 12 vii, viii, 15, 70, 71, 77, 81, 82, 93, 104, 151, 156, 163, 175, 177, 180, 182, 187, 195, 203, 208, 232, 233, 241, 242, 243, 248, 249, 250, 256, 257, 265, 267, 273, 290, 292, 294, 295, 296, 299, 311, 316, 317, 323, 324, 327, 337, 338, 341, 342, 344, 345, 346, 358, 373, 375, 376, 378

ABOUT THE AUTHOR

Shanée was born and educated in South Africa, graduating from the University of Cape Town in 1993 with a Bachelor of Business Science Honours degree in the special field of clinical psychology. She then pursued a career in the corporate world, first in planning, and later (after a year in London) heading up a Global Centre of Excellence, sitting on the board of directors, managing a local team, and training and consulting to a network of partners and clients in approximately 100 countries across five continents. In 2011, she moved to Sydney with her husband and two children, where she still lives. She has subsequently volunteered as a high school mentor through the Raise Foundation.

If you would like to contact Shanee for further information or to provide feedback, you can contact her via email at:
info@smarterstudysolutions.com.au
or via her website at:
smarterstudysolutions.com.au